Tasmanian Tiger
A lesson to be learnt

Eric Guiler & Philippe Godard

with the contribution of
David Maguire

Abrolhos
Publishing

Design : DEMAIN - Nouméa, New Caledonia
Typesetting and colour separations : VUE D'AILLEURS - Nouméa, New Caledonia
Printing : Stan Graphics Consultant Pte Ltd / Classics Offset
Printed in 1998

ISBN 0958 579105
© 1998 ABROLHOS PUBLISHING PTY. LTD
QVI Building. 25th level. 250 St George's Terrace. Perth 6000
Western Australia

*"To Lalage Guiler, dedicated assistant
of her husband Eric, who devoted
part of her life to the research of the tiger."*

Thylacine

They'll not find him in the hills
Above the slow fern gully;
He's gone to earth like a sunken creek
In an unknown valley;

Nor find the fur on the bent thorn
Nor hear him moan at the raw moon:
He stalks down the valleys of the years
With his old love, his old pain.

They'll not find him in the hills:
He's gone to earth in an unknown valley
With legends of coal and time in stone,
With the sly fern, with the gully.

from "The Other Meaning" poems by Vivian Smith

Contents

Foreword		8
Chapter 1	— A remarkable species	10
Chapter 2	— Earliest traces	28
Chapter 3	— Anatomy: a closer look	46
Chapter 4	— Tasmania: Thylacine's last home	58
Chapter 5	— The first eye witnesses	76
Chapter 6	— Thylacine and the artists	86
Chapter 7	— Persecution and bounty hunting	110
Chapter 8	— Tales of the tiger	142
Chapter 9	— In captivity	166
Chapter 10	— Expeditions and searches	178
Chapter 11	— The impossible dream	216
Appendices		229
Illustration credits		237
Bibliography		241
Index		249
Acknowledgments		255

Foreword

In all honesty, the authors must admit to being of those who no longer really believe - an attitude that certainly earns them no compliments - in the survival of the Tasmanian tiger, when others continue to nurture the hope of seeing the animal reappear one day. The latter argue that all the sightings reported in recent years, by various witnesses whose detailed accounts are sometimes troubling, simply could not be hallucinations nor the mere desire to see the spotlights of media shining upon them.

There is no question that, in the former Van Diemen's Land, there are practically deserted regions in which the animal could have found refuge; this would, however, be surprising, for some sixty years after the last specimen died in captivity at Hobart zoo, not one Tasmanian tiger has been found squashed on an isolated road, nor are there any unquestionable sightings of a survivor alive and well in the wild.

The word unquestionable is most appropriate, for none of the alleged sightings reported in over half a century by persons who claim to have come face to face with the animal along some track, are sufficient to really convince objective observers. But man is such that he needs dreams, mystery and marvel, and there will always be people who will swear by all gods that they have seen the yeti in the Himalayan snows, a giant sea snake in some faraway corner of the ocean, or the legendary Loch Ness monster.

Even the least sensational of newspapers willingly play the game, spreading the word of such supposed miracles. Did some of them not recently report the terror of an isolated African population that claimed to have seen a gigantic, unknown animal whose description was that of… a dinosaur! All this to say that the gap between fiction and fact, although insurmountable, is often crossed with great ease.

The example we have just given of the imaginary reappearance of an animal that disappeared a few million years ago is perhaps not such a good one. If the Tasmanian tiger were rediscovered one day, there would indeed be no reason to cry "miracle", nor to defy the theory of evolution, given that a number of our contemporaries have had the privilege of seeing a live Tasmanian tiger. At the very most, it would be a sensational occurrence that would, without doubt, make headlines around the planet.

The "tiger", curiously, is one of those animals whose very name arouses great interest throughout the world, and it is indeed perfectly natural to ask why it is so famous. A first answer is most certainly the choice of the animal's name, a most unfortunate one at that. The thylacine - as the Tasmanian tiger is now known to the world's scientific community - is a marsupial that, apart from the stripes on its fur, has absolutely no morphological resemblance to the great cat of Asia.

There is also the fact that the man in the street, be it in Sydney, Paris, New York or Rio de Janeiro, cannot imagine that a marsupial would demonstrate such aggressive instincts, feasting on flesh and blood. This belief is based on the concept that marsupials are friendly animals that we would all love to hold and cuddle. The word marsupial brings to mind the image of an animated soft toy, a koala, a wombat or to a lesser degree a kangaroo, sheltered in its mother's pouch, inspiring sympathy as it pokes its nose "out the window" to take in the first fearful impressions of the outside world.

It is beyond the realms of human understanding that the female of such an unquestionably cruel species as the thylacine, capable of slitting a lamb's throat with one strike of the teeth and licking up the still warm blood, would have the characteristic pouch of all marsupials, in which she can shelter up to four small balls of fur. Other carnivorous marsupials exist, feeding on insects, dead animals and all manner of small creatures, but the Tasmanian tiger would be the only marsupial strong enough to kill an active adult kangaroo.

But now it is time to get to the heart of the matter and relate, in the first instance, the circumstances in which this most remarkable of creatures came to claim its place among the fascinating wildlife of the Antipodes…

Chapter One
A remarka

ble species

The Tasmanian tiger is a marsupial and member of the *Dasyuridae* family. Marsupials are a widespread group of mammals, not only found in Australia, but also in both North and South America, as well as New Guinea. The Tasmanian tiger's closest relative is the Tasmanian devil and the two species of quoll, *Dasyurus maculatus* and *D. viverrinus*. It has been shown by DNA analysis that these three species separated from the main marsupial evolutionary line long after the separation of Australia from Antarctica and South America. Thus the thylacine is not related in any direct way to the South American marsupials.

All marsupials are distinguished by the fact that they give birth to very immature young that are then carried around in a pouch until such time as the off-spring are at a stage of development where they can run with the mother, and later roam on their own. A major disadvantage to this method of reproduction is that the young are vulnerable to predators when not in the pouch and away from the mother's direct supervision. Many other features differentiate marsupials from other mammals, and these are discussed later.

Although many others were used, the scientific name for the Tasmanian tiger is *Thylacinus cynocephalus* (animal with pouch and a dog's head). The other names result from the rarity of the animal and the fact that scientists were unable to compare their material with that of other researchers. Furthermore, taxonomic zoologists often change their minds about the proper name of the animal[1].

Tasmanian tiger is the most commonly used name, although this hasn't always been the case. Over the years the names wolf, Tasmanian wolf, and to a lesser extent thylacine have been used, but have never been common. The bushmen of yesteryear never referred to the animal as anything but "tiger" or "hyena". In this work, we shall use "Tasmanian tiger" and "thylacine" a word first used by Krefft in 1868. The term "slut", which refers to the female, must also be added to this list. We are not sure how widely the term was employed, but it was certainly in use in the central highlands up until recent times.

George Harris was the first to provide a scientific description of thylacine in 1808, published just five years after settlement began in the new colony. Harris, of whom no portrait has unfortunately become available to us, was the Surveyor-General and it shows remarkable activity on his part to bother with such matters while surveying the land allocations. It may also be an indication of the rarity of the species even then.

He was a talented man, practising law in Exeter (U.K), before becoming a surveyor and coming to Australia. He was a competent artist, and some of his watercolour landscapes of Australia were later copied by painters. Harris had a Quaker family background, and this may have started his interest in natural history, an interest that was stimulated by the discoveries of Banks and the learned Societies, and further enhanced by the new animals he saw. Harris also made several collections on his way to Australia.

Although sightings of tigers were reported in 1805, it was three years before Harris' first scientific description appeared. In handwritten manuscript, this was sent to Sir Joseph Banks along with a covering letter, then published by the Linnean Society of London in the same year. The animal was called *Didelphis cynocephalus*.

Harris' drawing of the Tasmanian tiger was the very first to appear in literature. Several features of the illustration are noteworthy - the number of stripes on the back is accurate but the claws on the rear foot are too long. The text states that the species is little known and very rare "*only two specimens ... having yet been taken*". Harris goes on to note that "*it inhabits amongst caverns and rocks in the deep and almost impenetrable glens in the neighbourhood of the highest mountainous parts*". This last comment underlines the possibility that, at the time, the Tasmanian tiger was at the low point of a population cycle, or perhaps that Aboriginal persecution over many centuries had driven the animals up into the higher country, with the introduction of sheep bringing them back down into more open regions.

Harris presumably saw living tigers as well as dead specimens during his work in the bush. Communications being as slow as they were at the time, it is not surprising that there was a three year interval between the first reported sightings and the description in zoological literature. Although very accurate, Harris' description gave rise to some misconceptions in later years. He noted that the tail was much compressed, but this is not shown in the illustration.

The thylacine shows remarkable similarities with the more commonly known dog or wolf, with a typically large head, evenly balanced backbone, legs of approximately equal length and hind legs with the femoral segment pointing obliquely forwards. Like the dog, it has a deep chest and non-retractile claws, running on its toes and not the whole foot. Despite these convergences in external form, the thylacine remains very clearly a member of the marsupial family in all respects. The similarity of external form is found in the skeleton, showing the same depth of chest, long limbs and the general features of a cursorial animal.

Previous page and left: Original drawing of the Tasmanian tiger by Harris in Trans. Linn. Soc. Lond. This very first representation of the Tasmanian tiger dates back to 1808. We owe it to Harris, and it accompanied, as we have seen, the description that he drew up for the Linnean Society of London. It seems probable that the author only had the remains of a thylacine to work from, and perhaps only even the hide of an animal skinned immediately after it was killed by a hunter. Indeed, apart from the stripes on the fur, nothing in this reconstitution reflects reality. Rather, due to the crouched body position and particularly the expression of the massive, thick-lipped head, it looks like some sort of bulldog.

French naturalist Geoffroy Saint-Hilaire turned his attention to marsupials in 1810 and realised that the genus *Didelphis*, to which Harris had assigned the Tasmanian tiger, was not suited to the species. He therefore placed it in *Dasyurus*, which he had established to cope with the flow of new species of carnivorous marsupials being identified on the new continent. The genus *Thylacinus* was established by Temminck in 1824 when

> *G. P. Harris to Sir Joseph Banks*
>
> > Hobart Town, River Derwent,
> > Van Diemen's Land Augt. 31, 1806
>
> Sir,
>
> I take the liberty of transmitting to you drawings & descriptions from the life of two animals of the Genus Didelphus, natives of this Country, which I believe are in every respect new, at least I have [not] seen any description of either.
>
> As I believe it is not uncommon for accounts of newly discovered animals to be communicated to the Royal & Linnean Societies If you, Sir, judge those sent worthy that Honor I shall be amply repaid for my labours -
>
> I am now preparing a work for the press under the title "Illustrations of the Zoology of Van Diemen's Land" - comprising accurate coloured drawings - from life, of Birds, quadrupeds, Fish, Insects etc etc. many of which are nondescript. I have already completed 150 drawings, and had I not been disappointed in receiving paper & drawing materials from England should have submitted specimens of the work by this opportunity for your approbation
>
> By the next Ship [*March 20 1807 inserted by G.P.H.*] I hope to do so with request that if you, Sir, think them worthy, you will permit the Honor of dedicating the work to your patronage -[23]
>
> > I remain, Sir,
> > With the utmost respect
> > Your most obedient & very hble servt.
> > Geo: Prideaux Harris
> > Dty. Surveyor General, & one of
> > H.M. Justices of the Peace, for the
> > Island of Van Diemen's Land etc etc etc
>
> P.S If there are any particular curiosities either in plants, animals etc which you would wish to receive from this country, I shall always feel a pride in executing the commands of Sir Joseph Banks - any communications to that effect, directed as above & left with my brother, H. B. Harris Esq. War Office will be duly forwarded.

Letter from G.P. Harris to Sir Joseph Banks.

he moved the species out of *Dasyurus* because he believed that it was sufficiently different to warrant the change. He also gave it a new minor name, *harrisii*, which was not accepted. The species had been quite clearly defined and, apart from some fussing about in early times, has suffered little from the attentions of taxonomists. Some new names were proposed, but all have been dropped under the rules of priority.

A new species was established in 1868 (Krefft) to cater for what was believed to be a new species of Tasmanian tiger, with a shorter skull than *T. cynocephalus*, which Krefft described as being only 17.5cm long. Also, there was no notch between the second and third premolar teeth, the second and third molars in each jaw were prominent and the canines were thicker than *T. cynocephalus*. He went on to name this species *T. breviceps* because of the short skull and called it the "bulldog tiger"; he rechristened *T. cynocephalus* as the "greyhound tiger". Allport almost immediately corrected Krefft's views and pointed out that the so-called bulldog tiger with the short skull was in fact the female thylacine, whilst the greyhound tiger was the male of the species.

Table of the different names for the tiger through the years.

Common name	Scientific name	Author	Year
Tyger		Knopwood	1805
Zebra opossum Zebra wolf	Didelphis cynocephala	Harris	1808
	Didelphis cynocephalus	Geoffrey	1810
Hyæna		Jeffreys	1820
Opossum hyæna		Evans	1822
	Thylacinus harrisii	Temminck	1824
	Peracyon	Gray	1825
	Peracyon	Griffith, Smith & Pidgeon	1827
hyæna Opossum		Mudie	1829
Tiger		Widowson	1829
Dog-faced dasyurus		Mudie	1829
Native hyæna		Ross	1830
	Lycaon	Wagler	1830
	Dasyurus leucocephalus	Grant	1831
Van Diemen land tiger (V.D.L. tiger)		Henderson	1832
	Thilacinus striatus	Warlow	1838
	Peralopex	Glober	1841
Tasmanian tiger		Backhouse	1843
	Thylacinus spelaeus (fossil)	Owen	1845
Dog-headed opossum Tiger wolf Zebra		Angas	1862
Bulldog tiger (female) Greyhound tiger (male)	Thylacinus breviceps	Krefft	1868
Thylacine		Krefft	1868
	Thylacinus major (fossil)	Owen	1877
Striped wolf		Wright	1892
Tasmanian dingo		Lyne	1886
	Thylacinus rostralis	de Vis	1894
Tasmanian wolf		Anderson	1905
Marsupial wolf		Lucas & Le Souef	1909
	Thylacinus potens (fossil)	Woodburne	1967

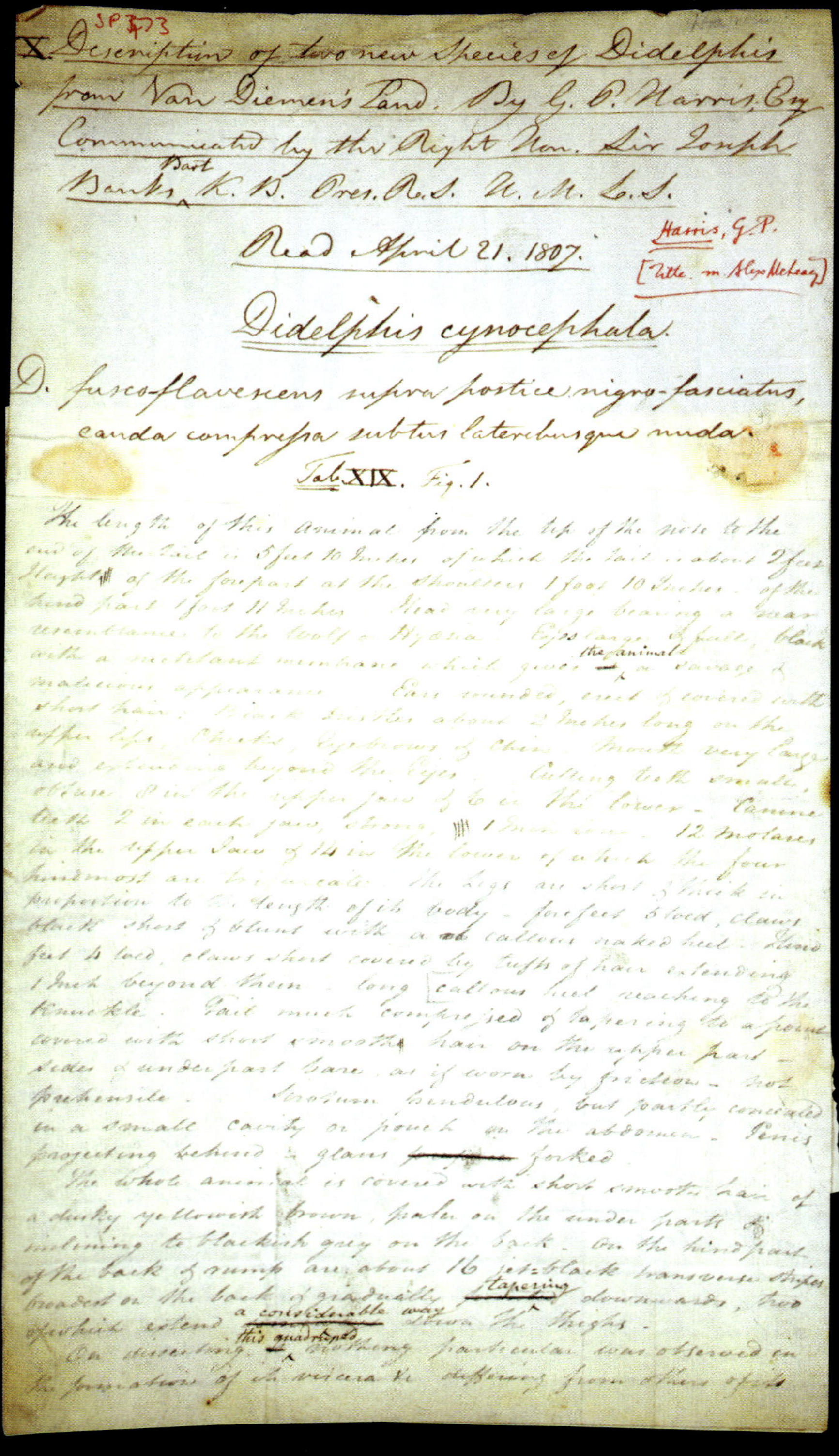

CHAPTER ONE — A REMARKABLE SPECIES

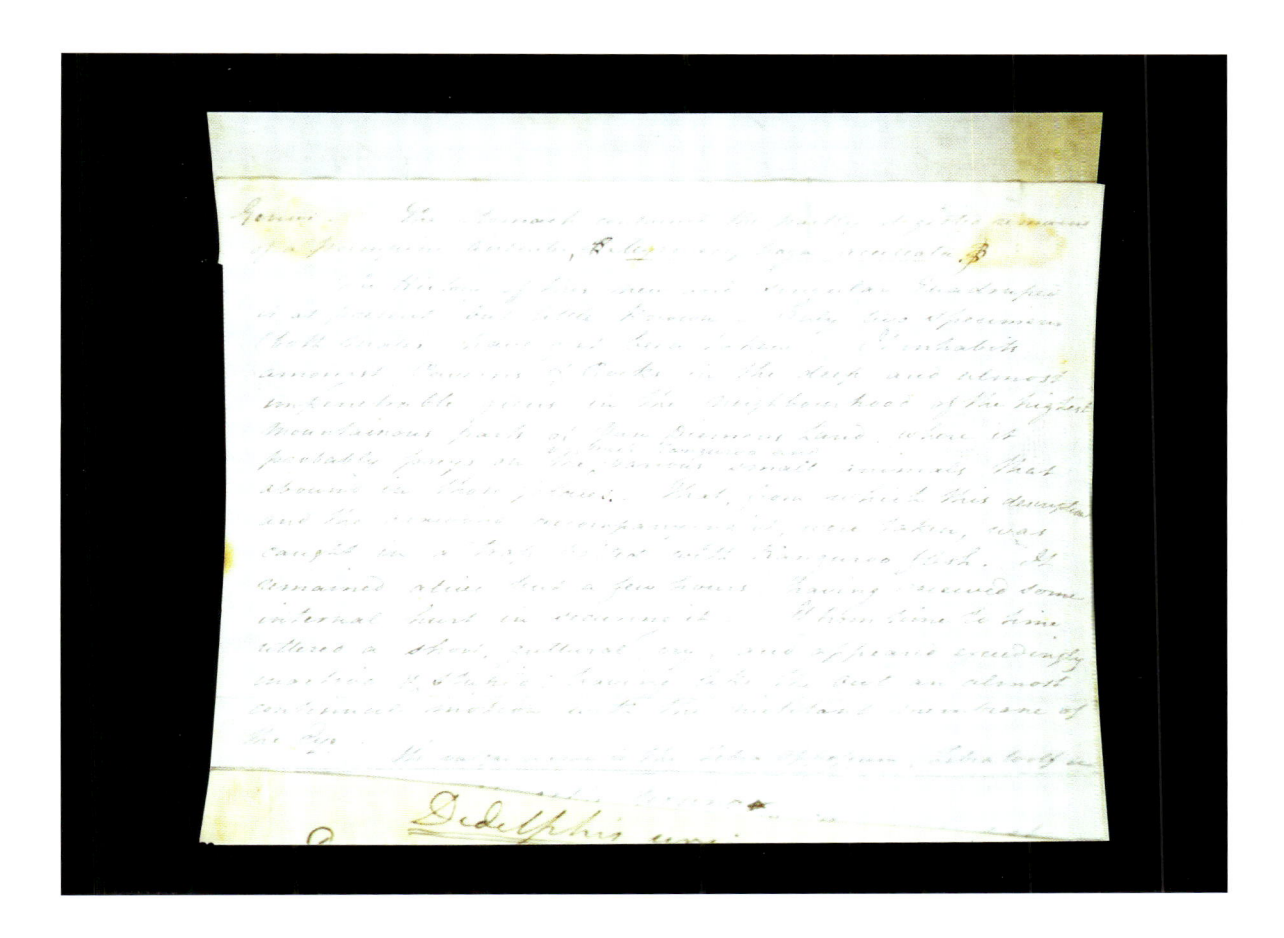

Krefft had in effect established a new species to differentiate between male and female *T. cynocephalus*.

Left and above:
Harris' handwritten description of the Tasmanian tiger.

Pathetically little is known of the biology of the Tasmanian tiger. It is possible to make reasonable deductions from the early anatomical descriptions, however the overall picture can only be completed from the sketchy information available, notwithstanding some contradictory statements and other obviously incorrect observations.

The lack of investigation of a species that is both interesting and rare enough to excite curiosity even in early times must be considered in context. The Tasmanian Museum and the Queen Victoria Museum were the only local research institutions in the days when tigers were still relatively abundant. Both museums had very few staff and employed no trained zoologists. As was typical of institutions in the post-Darwin period, both museums were more concerned with amassing collections of animals to be preserved, classified and then put on display for the public. The Foundation Chair of Zoology at the University of Tasmania was only established in 1909, by which time the decline of the thylacine was well underway and specimens were hard to obtain.

> XI. *Description of two new Species of Didelphis from Van Diemen's Land.* By G. P. Harris, Esq. Communicated by the Right Honourable Sir Joseph Banks, Bart. K.B. Pres. R.S. H.M.L.S.
>
> Read April 21, 1807.
>
> DIDELPHIS CYNOCEPHALA.
>
> DIDELPHIS fusco-flavescens supra postice nigro-fasciata, caudâ compressâ subtus lateribusque nudâ.
>
> TAB. XIX. Fig. 1.
>
> The length of this animal from the tip of the nose to the end of the tail is 5 feet 10 inches, of which the tail is about 2 feet. Height of the fore part at the shoulders 1 foot 10 inches—of the hind part 1 foot 11 inches. Head very large, bearing a near resemblance to the wolf or hyæna. Eyes large and full, black, with a nictitant membrane, which gives the animal a savage and malicious appearance. Ears rounded, erect, and covered with short hair. Black bristles about 2 inches long on the upper lips, cheeks, eyebrows, and chin. Mouth very large, and extending beyond the eyes. Cutting teeth small, obtuse, 8 in the upper jaw, and 6 in the lower. Canine teeth 2 in each jaw, strong, 1 inch long. Twelve molares in the upper jaw and 14 in the lower, of which the four hindmost are trifurcate. The legs are short and thick in proportion to the length of its body. Fore feet 5-toed, claws black, short, and blunt, with a callous naked heel. Hind feet 4-toed, claws short, covered by tufts of hair extending 1 inch beyond them, with a long

Description of the Tasmanian tiger Trans. Linn. Soc. Lond. IX, 1808, 174.

At the beginning of the 20th century, overseas research institutions were advertising for tigers, but their requests could seldom be satisfied. The tiger was already scarce, and trappers could probably collect more money for a carcass by offering it for sale on various properties. No-one ever contemplated field research programmes, and the thylacine had almost disappeared by the time the science of ecology started developing in Tasmania. Nevertheless, there are records of limited numbers of the species, even as late as 1930, finding their way to museums, zoos and the University of Tasmania. Some criticism of these and other institutions is justifiable, given that it was already obvious that the tiger population had drastically reduced and that it no longer existed on the Australian mainland.

Over the years, a zoo's role in the community has changed. As well as displaying animals to the public, the modern zoo is now concerned with the conservation of species and reestablishing their former habits. A shame that this change came too late for our Tasmanian tiger.

There is no evidence of thylacines being observed mating in captivity or in the wild. By 1860, it appeared in literature that the tiger retired to mountainous and

CHAPTER ONE — A REMARKABLE SPECIES

> *Mr. HARRIS's Description of two new Species of Didelphis.* 175
>
> long callous heel, reaching to the knuckle. Tail much compressed, and tapering to a point, covered with short smooth hair on the upper part; sides and under part bare, as if worn by friction; not prehensile. Scrotum pendulous, but partly concealed in a small cavity or pouch in the abdomen. Penis projecting behind: glans forked.
>
> The whole animal is covered with short smooth hair of a dusky yellowish brown, paler on the under parts, and inclining to blackish gray on the back. On the hind part of the back and rump are about 16 jet-black transverse stripes, broadest on the back, and gradually tapering downwards, two of which extend a considerable way down the thighs.
>
> On dissecting this quadruped, nothing particular was observed in the formation of its viscera, &c., differing from others of its genus. The stomach contained the partly digested remains of a porcupine ant-eater, *Myrmecophaga aculeata*.
>
> The history of this new and singular quadruped is at present but little known. Only two specimens (both males) have yet been taken. It inhabits amongst caverns and rocks in the deep and almost impenetrable glens in the neighbourhood of the highest mountainous parts of Van Diemen's Land, where it probably preys on the brush Kangaroo, and various small animals that abound in those places. That from which this description and the drawing accompanying it were taken, was caught in a trap baited with kangaroo flesh. It remained alive but a few hours, having received some internal hurt in securing it. It from time to time uttered a short guttural cry, and appeared exceedingly inactive and stupid; having, like the owl, an almost continual motion with the nictitant membrane of the eye.
>
> It is vulgarly called the *Zebra Opossum, Zebra Wolf*, &c.
>
> DIDELPHIS

inaccessible regions to breed, and that they were uncommon in settled parts of Tasmania. There is no evidence of this migration later in the century, when tigers were plentiful in pastoral areas. As for the other two Dasyures, the native cat and the Tasmanian devil, neither moves away from the home range to breed.

In 1928, the director of the Tasmanian Museum, C. E. Lord, stated that thylacines laid a scent trail across the countryside during the breeding season. Although some marking may occur, there are no descriptions of any glands that would serve this purpose. Several authors have claimed that the Tasmanian tiger has a lair or some other retreat, variously described as a hollow log or tree, a cave or a rock cavity, where the young are reared. There is no doubt that the tiger has favourite sleeping places, but this is very different from establishing a den for bringing up the young. Many old trappers stated that thylacines did not have a permanent lair.

The number of young carried around in the pouch varies from one to four, which is the maximum number of nipples available. The lack of reference material makes it impossible to calculate an average number of pouch young per female, but it can be assumed that thylacine is polyovular, which is the usual Dasyure pattern. The number of eggs produced by the ovary in any one season is not known, but observations of the native cat and the Tasmanian devil show that many more eggs are produced into the uterus than are required to fill the available pouch space. Hughes found up to 100 fertilised eggs in the uterus of a Tasmanian devil, but the fate of these has yet to be determined.

It is interesting to speculate whether all eggs reach full term or are resorbed at some time prior to full development. If all reach term and are born, then there would be an extremely high mortality rate in the struggle for the four nipples. It is more likely that a female would successfully give birth to two young in her first breeding year, three or four in the peak years, followed by a decline as the animal grows old. The gestation period is not known, but if it follows the usual Dasyurid pattern, it should be about thirty-five days.

The Lands Department account books are an interesting source from which some breeding information has been gleaned. The bounty for post-pouch young animals was ten shillings, and a total of 150 pre-adult animals were submitted for bounty. Clerks differentiated between pups and half-grown young in about two-thirds of the bounty claims. A majority of the pups were found in the winter months, about one month before the half-young appeared.

Breeding appears to take place all-year round, with pups and half-young presented for bounty in every month. This is further substantiated by H. Pearce, an old-timer who remarked that young could be found in pouches at all times. The Dasyures all tend to have a restricted breeding season, the Tasmanian devil only mates in the March-April period, which is in tune with the young assuming a free life at the most advantageous time, in late spring or early summer, whenever food and mild weather provide the best chance to become established. The reproductive organs regress in the the non-breeding season, making reproduction impossible, although some cases of out-of-season breeding have been recorded in the Tasmanian devil.

As for the Tasmanian tiger, it is assumed from all accounts that only one litter is carried each year, which restricts the the species' capacity to recover rapidly from a catastrophe or excessive hunting pressure. The female is receptive or "on heat" once a year, there is no evidence of retarded or suspended development of the embryos, which is to be expected, being a feature of only some kangaroos. The major features of thylacine breeding biology resemble the other Dasyures, with a restricted breeding season and some out-of-season reproduction. Mating probably occurs in December or thereabouts, which means the young are born in January, reaching the pup or half-grown stage by June-September as shown in

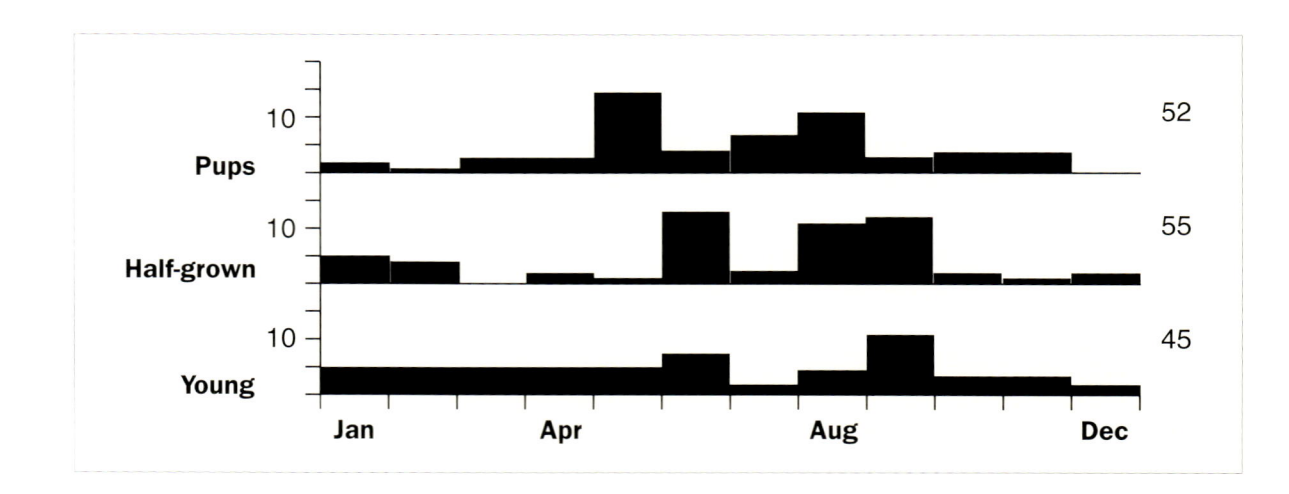

Monthly distribution of young Tasmanian tigers caught during the bounty period 1888-1909. The young are in three groups based upon their size. The «young» group contains animals which were insufficiently described to be assigned to either of the other two groups.

the monthly distribution figure, and independent life in early summer.

It has been claimed that the young spent a three month period in the pouch, but this seems very short in comparison to many other marsupial species, which are fully furred with all systems functional at 100 to 110 days. At this stage, the advanced embryo has become a baby mammal, but is still not capable of assuming an independent existence. By about 140 days the Tasmanian devil is very mobile and capable of running about, although still nursed by the mother. Assuming that the breeding biology of the Tasmanian tiger is similar to that of the devil, a pouch life of 130 - 140 days would be expected, followed by another month or so in which the young are hidden while the mother hunts. During this time the young would still be largely dependent on mother's milk, until the next phase when they run after the mother, a sign that the weaning process is underway.

In 1945, a 7.5 mm crown to rump long pouch young was described as naked, devoid of colour pattern and showing no development of fur. A 288 mm long individual was furred, with open eyes, and would probably have been about full term.

Several old timers, including H. Pearce, have stated that the young run with the mother upon leaving the pouch, hiding in shelters while the mother hunts, and then follow the hunting trail at a later stage. The mother is said to move her family night after night, all members hiding in any convenient spot by day. Although contradictory to the beliefs of many authors, this practice is substantiated by Mrs Louisa Ann Meredith (1881) who lived on the east coast of Tasmania and was in a position to gather first-hand information from trappers who, like Pearce, were experienced in the ways of the tiger.

Three young thylacines in their mother's pouch.

Pearce's theory is supported by the number of female tigers caught with cubs in tow. If the cubs were left in the lair while the mother hunted, they would not have been taken when the mother was captured. On the other hand, if the young run with the mother, it is likely they would stay with her when she was snared. As early as 1852, it was said that the tigers formed family units, and several trappers said that the young ran with the parents for a time, and that "packs" of up to six individuals had been seen.

The young may stay with the parents for some months, possibly until the following breeding season. The fact that young were seen with both parents is unusual, the formation of a family unit, however temporary, implies that a pair bond would exist between the parents. If Tasmanian tigers indeed formed this bond, it would be unfavourable to the rehabilitation of the species, and be the cause of a slow recovery due to the difficulty in finding a new mate in an already scarce population.

There are no direct observations of thylacine movements. It is not known if they are sedentary, have a home range or territory, or even if they are vagrants, however some evidence obtained from the Woolnorth station diaries and old trappers gives clues as to the thylacine's habits. Tiger hunts are described in the diaries and the evidence quite strongly suggests that the animal made little attempt to move away from an area even in the face of persecution. This implies thylacine has a home range, if not a defined territory and is a view supported by A. Youd of Deloraine, who trapped in the Lake Adelaide - Golden Valley region, and said that *"once you found where they lived then all you had to do was to stick at it until you caught them"*. A similar assumption was suggested by Wilf Batty in his account of the shooting of a thylacine in Mawbanna in 1930, when he made the point that the animal had been in the area for some time prior to the shooting. This also supports the home range theory.

Right:
The National Zoological Park in Washington had the privilege of accomodating a female thylacine and her three off-spring at the beginning of the 20th century. This gouache painting, completed in 1902 in a real-life situation by an artist named Gleeson, is an extraordinary document`in so far as it is the only illustration of its type. Although none of the baby animals can be seen in the marsupial pouch, the pouch does appear stretched, as if small off-spring were in residence.

What may appear to be a contrary view was expressed by George Wainwright, the last "tiger man" on Woolnorth station. Wainwright believed the tigers moved up and down the coast feeding on the abundant game living in this excellent habitat. The tiger man lived at Mt Cameron West, catching most of his Tasmanian tigers on the coastal runs. Wainwright believed he caught them as they were moving along the coast. It is possible that as a result of tigers being caught and removed from the coastal runs, other animals would move from sparser areas to this lush habitat, thus giving an impression of constant migratory movement.

The bounty records enable identification of the months in which the animals were killed, with more taken in winter than in summer. From 1888 to 1908, about seventy tigers were caught every month from November to March throughout the state, increasing to over one hundred between May and October, peaking in July-August. The peak of kills in winter is mainly related to the seasonal nature of snarers' activities, although there is some evidence from Woolnorth that indicates that tigers were more active in the winter months. Towards the end of the 19th century, the fur industry boomed, and many highland shepherds turned to snaring whenever the sheep were returned to the lower winter pastures. The high price paid for winter skins would have been a powerful financial incentive.

Some old records refer to tigers moving down from the mountains in winter and returning in late spring and summer for breeding, but there is too little evidence to corroborate this view, and none of the other Tasmanian mammals undertake such migration. The seasonal nature of catches led James Malley of Trowutta to postulate that thylacine moves to the coast in winter, benefiting from the warmer environment and better breeding conditions. He also suggests that the tigers traditionally moved up the coast, but that in later times agricultural activity in coastal regions forced thylacines to move through the thick inland forests. It would be most unusual for a mammal to alter a traditional migratory route, and there is no evidence that seasonal migrations in fact took place. Although Malley's suggestion does not fit in with the idea of a home range, his view is nonetheless interesting with so little known about the animal in its wild state.

It is generally agreed that Tasmanian tigers are nocturnally active, despite 47% of all sightings being recorded in daylight hours. Most of the species upon which thylacine feeds become active in the bush in the early evening, grooming themselves and moving to open areas to nibble at the grass, and some are still to be seen lazing in the sun and dozing well after sunrise. There is no reason to assume that the tigers would not be active at these times, and Sawley, who lived in north-west Tasmania between 1900 and 1930, firmly stated that thylacines always commenced hunting just on dusk.

TASMANIAN TIGER — A LESSON TO BE LEARNT

Learning to hunt.

In 1938 the observation was made that the Tasmanian tiger in New York Zoo spent most of the day basking in the sun, and that the animal had poor vision in sunlight. Captive animal behaviour is not necessarily the same as that of animals in the wild, but wild tigers have also been observed basking in the sun. The claim that they have poor day-time vision is probably incorrect, and not supported by any other writers.

Any predator has to get food by hunting and catching prey. In this respect, the thylacine is faced with the problem of catching prey that is sometimes larger and faster than itself. Because it is a rather slow runner, the tiger either has to stalk prey and pounce like a leopard, or else hunt in a pack and use co-operative hunting techniques, in the manner of wolves. There is no evidence that the tigers did catch prey by pack hunting.

The similarity in appearance to wolves and dogs suggests that the Tasmanian tiger is a fast runner, but all the old trappers were unanimous in saying that the thylacine did not have enough speed to run down prey in a straight chase. They told of two hunting

CHAPTER ONE — A REMARKABLE SPECIES

techniques used by the thylacine. It is known that a wallaby under chase will run in a wide circle, and that a thylacine would cut across the circle to grab the wallaby as it went past. The thylacine is a persistent runner, loping after its prey until the animal finally collapses from exhaustion. These methods appear feasible and would be possible hunting techniques, but all trappers concur in saying that no co-operative hunting techniques were deployed.

The few valid accounts of hunting methods provide a picture of a tireless predator running down its prey, not sprinting, but in steady pursuit. Thylacine's deep, large-capacity chest is adapted to this hunting method, and is a parallel to canine hunters. Keast compared wolves' hunting methods with those of the thylacine and concluded that the wolf was more specialised in pursuit, but makes the point that the tiger was "*adequate for its role on the Australian continent*". One might add that the tiger would have quickly become extinct if it had not been so adapted, or had adopted a different lifestyle. Keast implies that some of the thylacine's food was stalked rather than pursued, but this is not compatible with the few available historical records. His study indicates that the tiger would be at a disadvantage in a competitive situation with the more efficient dogs. Moeller showed that the skull and limbs of the Tasmanian tiger show little adaptation to this method of hunting in comparison to the Tasmanian devil and native cats, despite having a relatively longer neck than these other species.

We have uncovered no written descriptions of the method thylacine uses to actually kill prey, but old-time hunters do seem to agree on the subject. Pearce told the author that "*they hunt by lying in wait for their prey then jump on it. Kangaroos are killed by standing on them and biting through the short rib into the body cavity and ripping the rib cage open*". Tasmanian tigers are anatomically suited to this method of killing, the extraordinarily wide gape enabling it to seize the neck or chest of a wallaby and crush it. This method was also described by Sawley, who stated that the liver and the heart of prey were eaten.

25

TASMANIAN TIGER — A LESSON TO BE LEARNT

This photo, of unknown origin or location, shows three thylacines. It is important because to our knowledge it is the only photograph showing a female with young in the pouch which is distended. None of the young are visible. It may be that it was taken in Adelaide Zoo about 1900. The pouch size suggests that the young may be of 70-90 days old.

Pearce's story must carry some weight, and although his description of hunting and killing is somewhat different from the general point of view, the "lying in wait" would fit in with the final phase of the chase after the pursuer had cut across the wide circle mentioned above. Pearce did agree that the tigers were slow and could not simply run prey down. A few trappers believed that one thylacine drove the prey towards its mate, which then jumped out of hiding and killed. This implies a degree of co-operative hunting technique that is not reported by other observers, although, like Pearce's account, it can be reconciled with cutting across the circle in the final phase of the hunt.

Prey is killed then eaten. Many trappers spoke of a particular liking for vascular tissue, to which Sawley added that "*the tail was eaten about two inches from the anus to provide roughage*". The heart, lungs, kidneys and liver are consumed together with some of the meat from the inside of the ham, but the remainder of the carcass was left, probably to be devoured by Tasmanian devils.

In January 1952, a dead wallaby that had been killed in this fashion was found about 10 kilometres north of the Pieman River. The corpse was still warm, which indicates that the tiger may have been disturbed while feeding. In 1976, about 16 kilometres from this site, a pademelon (small wallaby) was found dead with its throat eaten out. Like the earlier kill, only the organs mentioned above had been eaten and there was no sign of other bites or even of a struggle. This killing method is quite different from that used by dogs, which typically worry their prey. Tiger cats enter through the anus or the belly, whereas devils are not selective and will eat any convenient part of injured or dead animals.

It is generally agreed that Tasmanian tigers kill their own food in the wild, and never return to the site of a kill. Sheep carcasses have been injected with strychnine in the hope of poisoning tigers, but were never eaten by them. In captivity, the tigers ate quite different food, indeed almost anything that was offered to them. In 1850, Gunn kept some specimens in captivity and fed them mutton, however they preferred the parts containing bone, leaving the vascular tissues. Tigers held in captivity by Rowe in 1913 ate the bones as well as the meat of wallabies, whereas the London Zoo specimens were fed rabbits and caught pigeons. If they are to survive, captive animals must eat what is presented to them, and this provides no indication as to their feeding habits in the wild.

1 - cf Table of Synonyms p.15.

TASMANIAN TIGER — A LESSON TO BE LEARNT

Chapter Two
Earliest

traces

The Tasmanian tiger lived in many more places than it did until recently in Tasmania, despite being named after the island state. Fossil history of the species is not fully known, and it has not been identified earlier than the Miocene period, in which the diversification of primates, including the early apes, occurred. The actual origins and relationships of the Tasmanian tiger are obscure.

Since the beginning of the 1950's, fossil remains of thylacines have been identified in Victoria, South Australia and Western Australia, as well as Papua New Guinea, where a single specimen was discovered in 1963. Countless remnants have been found throughout Western Australia, but more particularly in the caves situated in the southern part of this immense state (Mammoth cave, Strong's cave, Devil's lair). In general, these remains comprised skeleton fragments, identified by the very specific shape of the jaw. Using these finds, comparisons were made, revealing that the species that inhabited this part of the continent was of a size somewhat smaller than the species that existed in Tasmania until recently. This discovery is obviously of great interest.

Western Australia can also lay claim to possessing an extraordinary mummified "tiger" specimen. It was discovered in October 1966 by Mr David Lowry and his wife Jacky. At the time, David was working as a geologist for the W.A. Geological Survey, and together with his young wife, was a keen explorer of caves. They convinced their superiors that if they wanted a geological study of the Nullarbor, they had better send someone who liked going down caves and hanging over cliffs, as this was the only way to see the layering in the rocks.

It was for this reason they undertook the stratigraphic exploration of one underground excavation, amongst many others, that was to become famous. In the words of the explorers themselves:

"The cave, known as Thylacine hole ever since our discovery, is located on Mundrabilla Station, about 12 miles north-west of the homestead, and the same distance north of the Eyre Highway. The site is part of the Hampton Tableland and lies midway between Madura (South Australia) and the tiny village of Eucla (Western Australia), where there was a Telegraph wire office in earlier times. Today the moving sand dunes are threatening to bury the office ruins. This cave was initially explored by a party from the Sydney University Speleological Society in 1964, but no zoologist was among them.

"The entry to Thylacine hole is marked by two small "willow" trees, on the boundary between a clay flat to the south and a stony ridge to the north. The access to the shaft, which is about 38 feet deep, is a circular hole 3 feet in diameter in a rock pavement at the bottom of a depression 25 across and 4 feet deep.

"At its base, the cave opens out in all directions to form a low roofed volume in which the floor follows the undulations of the roof.

General aspect of the inside of the cave in which the remains of six thylacines plus those of a lot of other marsupials were discovered by Mr and Mrs Lowry.

CHAPTER TWO — EARLIEST TRACES

Above: Mrs Jacky Lowry at the entrance of the famous Thylacine Hole (Nullarbor).
Left and next page: The mummified specimen at the place it was precisely occupying in the cave when discovered in October 1966.

"The cave is about 500 feet across and reaches a maximum depth of 90 feet in a spacious clay-floored chamber 240 feet west of the entrance. Exploration is usually limited by the floor and roof wedging together rather than by definite walls. Relatively abundant calcite formation as well as halite "flowers" and straw stalactites occur.

"While we were surveying the cave we discovered a well-preserved carcase of a thylacine (Tasmanian tiger) on top of a rock pile 450 feet from the entrance. The remains of many other animals were also found, including five other thylacines, a Tasmanian devil, native (marsupial) "cats", possums, bandicoots, a wombat, large kangaroos, wallabies, rats, bats, dogs (dingoes), a feral cat, barn and masked owls, an emu, kestrels and other smaller birds, snakes, goannas, a centipede, moths and cockroaches.

"Although *Thylacinus* has been recorded in caves of the Eucla Basin, this carcase was in a more complete state of preservation than any previously found. It was laying on its right side with skin and hair still intact on the exposed side.

"The characteristic dark bars could be seen on the back of the thylacine. Although thylacines closely resemble dogs, a simple character which can be used to distinguish them is the number of incisors in the upper jaws: thylacines have eight whereas dogs have six. A count of eight upper incisors therefore confirmed our discovery as a thylacine.

"Recognisable, desiccated tissue was still present, including the left eye ball and the tongue. The animals limbs were dis-articulated. Limb joints were very brittle, and broke to expose a dark tarry substance during the initial examination. The tail was found 12 inches away from the body, possibly moved there by rats which may also have chewed at the abdomen. The presence of a few pupal cases indicated that flies had destroyed some of the soft tissue.

"This material was collected in collaboration with the W.A. Museum. A plywood coffin and a lot of aluminium foil to pack the carcase of "Old Hairy" (such was the name we gave to the mummy) were brought to the site. Mr Duncan Merrilees, curator of Paleontology at the museum, showed great excitement. His immediate opinion was that, despite the belief that thylacines had died out thousands of years ago on the mainland of Australia, one was in the presence of an animal that had, in recent times, accidentally fallen in to the cave, from where it was unable to climb out.

"It then probably wandered around the cave until it died of thirst and starvation, or fell between the limestone blocks that cover the floor. This scientist was therefore ready to declare the Nullarbor a thylacine reserve!

Close-up view of the mummy. After more than 4,000 years, the state of conservation is astonishing. Most of the hair, teeth, tongue and even an eye-ball are present!

"Three radiocarbon date determinations have been made on hair and desiccated skin, muscle and other soft tissue which lay immediately beneath our specimen, and clearly represented part of the animal concerned, because no other source lay closer than about 100 feet away. As soon as the desiccated carcase was lifted, about 18 g of this loose hair and other dry tissue was transferred by knife blade to aluminium foil and closely wrapped. The dates (N.S.W. 28 a, b and c) on this material are 4,650 +/- 104, 4,550 +/- 112 and 4,650 +/- 153 years before present (B.P.) respectively. The sample for date N.S.W. 28 c was boiled with hydrochloric acid to remove geologically old carbon in carbonate derived from the roof and walls of the cave.

"Thus the Thylacine Hole specimen, despite its very complete preservation, is older than the other one found a little bit later in the Murra-el-elevyn Cave of this same Nullarbor and which remains the youngest dated thylacine from the Australian mainland.

"A musty odour of decomposition still surrounded the carcase when we found it. The presence of hair indicates that hair-eating beetle had not found the animal, possibly due to a combination of the animal's freshness and its location so far from the entrance.

General view of the mummy after recuperation by the scientists of the W.A. Museum. This specimen provided important information on the existence of the Tasmanian tiger on continental Australia.

"The problem was to know why "Old Hairy" was so well preserved. The mummifying is probably due to rapid desiccation in the warm dry air of the cave. The air in Nullarbor caves is certainly very dry (in October, the temperature was 20° C and the relative humidity 67%) and there is so little seepage through from the surface that what little there is evaporates in the pores of the limestone, wedging off grains of limestone as the sodium chloride concentrates and precipitates. Perhaps the sodium chloride crystals lying around in the cave floor deterred the beetles.

"Also of importance was the discovery of skeletons of five other thylacines in the same cave. All the skulls and teeth are virtually complete, and in two cases, the postcranial parts were also fairly complete. These remains have significantly increased the amount of material available for study of the mainland race of thylacines. The skeletons were found lying on the floor or in crevices, and we collected virtually all the valuable material from the cave. A stratified clay deposit in the lowest part of the cave was barren, and thus there is little scope for further collecting.

"Another problem was why so many animals (thylacines and others mentioned above) had gone down the hole and why such a proportion of them were carnivores. One theory I had was that if a kangaroo fell down by accident while bounding across the grassy plain, its carcase would fill the cave with the smell. There is a large diurnal fluctuation in air pressure and the caves act as a large reservoir connected to the atmosphere by a few small openings. These "blowholes"

CHAPTER TWO — EARLIEST TRACES

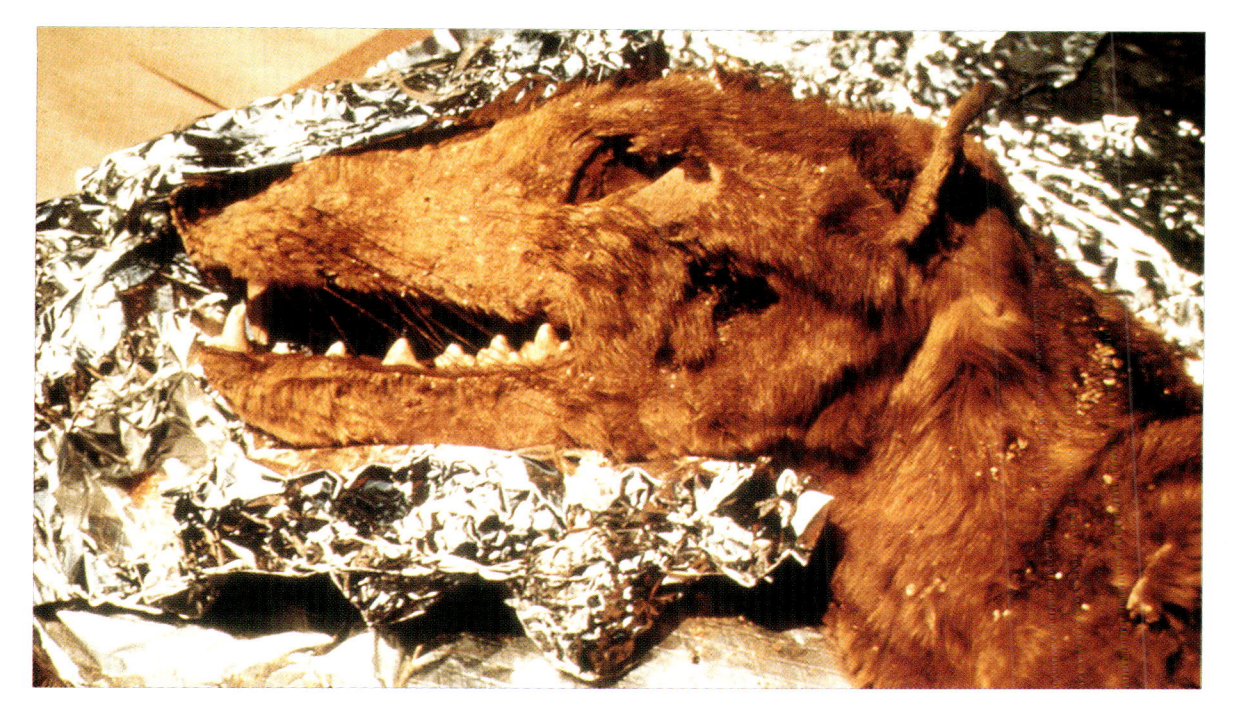

Detail of the head, whose length (in comparison to that of a dog) is particularly remarkable.

alternately suck and blow twice in 24 hours. Perhaps the smell of the carcase blowing out of the caves could attract scavengers to the edge of the blowhole from a long way down wind. But it is hard to see the animals being so stupid as to fall in accidentally.

"Compared with thylacine remains found in caves in south-western Australia, the Nullarbor specimen which we found is very small. Thylacines show quite marked sexual dimorphism, with females being smaller than males, not only in height, but also in the dimensions of their teeth. Jacky's work on "Old Hairy's" teeth showed that they were small compared with the skulls found in the cave. This suggests that the carcase may be that of a female."

The specimen in question was remarkable for the fact that it was almost completely intact - skeleton, hide, fur and whiskers. Even the tongue and the eyes were preserved! This rare find could well be explained by the absence of vermin at the site and by the very quick drying of the cadaver due, as previously noted by David Lowry, to the relatively high temperatures and extremely dry air in this specific underground environment. The unfortunate animal met its death at over 100 metres from the base of the chimney linking the cave to the outside world. It would therefore be logical to think that the animal did not perish when it fell down the chimney, and that it had wandered around for a time in the darkness looking for a way out, before giving up and dying from hunger and thirst.

The remains are part of the treasures of the Western Australia Museum of Natural History in Perth, where they are exhibited from time to time. Dating of the remains is

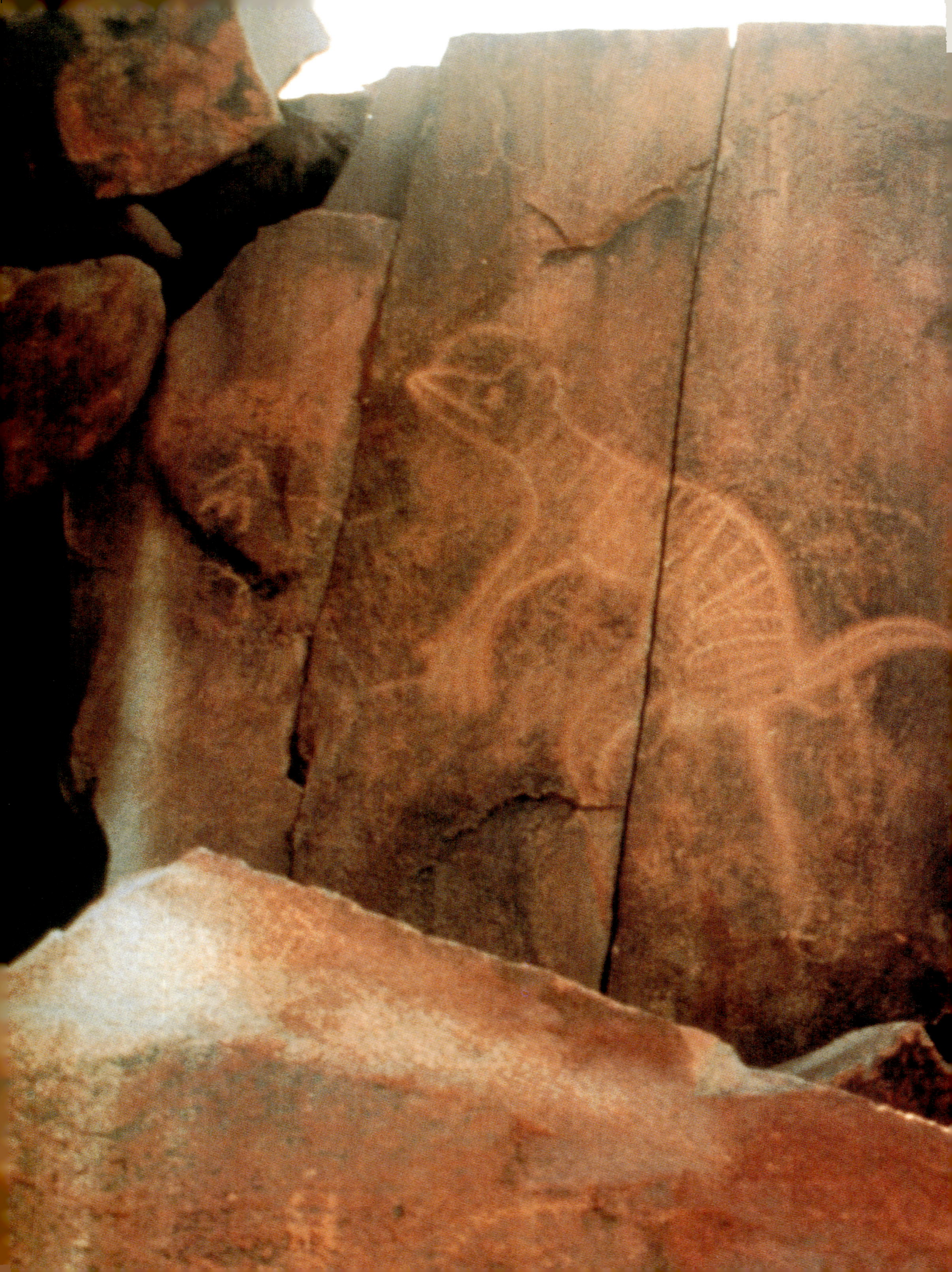

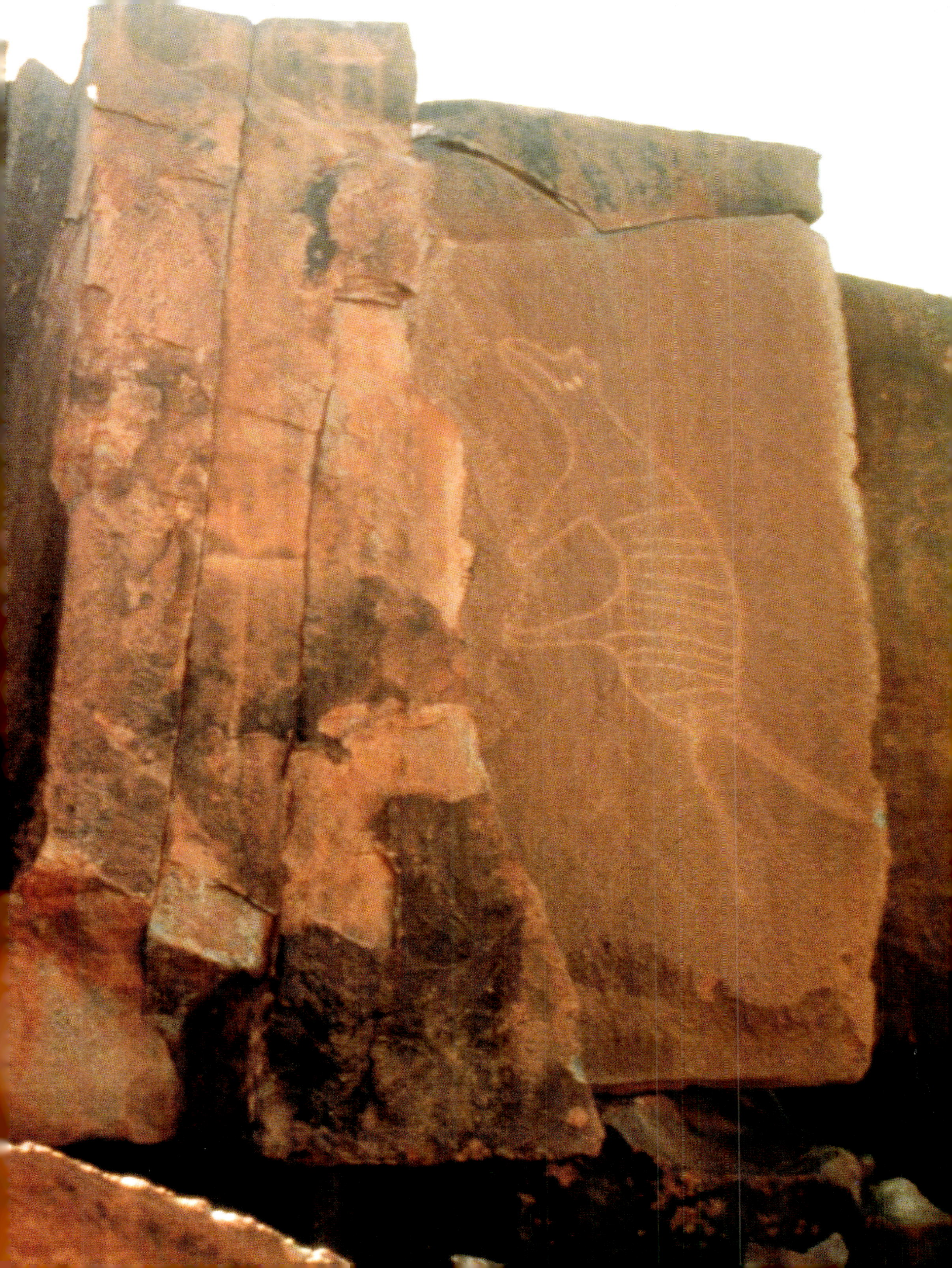

within a time frame going back 7,000 years, established from datings of other fragments collected in various places. Significant tiger and devil populations existed on the continent in this period.

The interesting question to ask is obviously when this population actually died out. For a long time, the "youngest" mummified carcases discovered in Western Australia were said to be about 3,000 years old. However, the recent discovery in the Kimberley Mountains of a pile of bones belonging to a variety of animal species, including the thylacine, has undermined the accuracy of this assumption. Carbon 14 datings in fact gave an age 0 ± 80 years before… 1950! A degree of uncertainty nonetheless exists, as it is still possible that the dating relates to some material that is foreign to the animal we are specifically interested in[1]. If this were not the case, and if thylacine did still exist in the Kimberleys around 1860, this would be amply sufficient evidence to reinforce the conviction of those people who, from time to time, claim to have encountered the Tasmanian tiger on the mainland, as others have done in Tasmania.

There are frequent claims of tiger sightings from all over Australia, mainly from Victoria and

Previous spread: **This faithful representation of the thylacine by an Aboriginal artist is found on Angel Island, Dampier Archipelago, Western Australia.**
Below: **Closer view of the same animal, a particularly spectacular illustration against such a dark background.**

CHAPTER TWO — EARLIEST TRACES

Another Aboriginal rock engraving on Angel Island.

the Western and Southern States, with some sightings reported in New South Wales and Queensland. Some of the reports have very accurate and credible descriptions, and there can be no doubt that the viewer believed he had seen a Tasmanian tiger. But there is a lack of hard, convincing evidence to back up the sighting claims. Cat, dog, fox and other unidentifiable footprints have been recorded, but no irrefutable proof has been discovered. If small pockets of tigers did still exist, then it is almost certain that Aborigines would know of their whereabouts.

Assuming that Tasmanian tigers have vanished from all of continental Australia, what then is the explanation for their disappearance? The climate about 6,000 years ago was more humid than today. It began drying out about 5,000 years ago, and has changed little in the last 3,000 years. The moist period would have suited thylacines better than today's drier conditions, but the tigers are known to have survived well into this arid period in continental Australia. We cannot conclude that the changes in climate at the time were responsible for the disappearance of both tigers and devils from the continent, and other factors such as competition amongst species must be considered.

Four species were possible competitors to thylacine, two of which were canine, the dog and the dingo, the third being human, and finally the Tasmanian devil. The dog is believed to have come to Australia with the Aborigines perhaps some 20,000 years ago, although a more recent date of about 8,600 ± 300 years ago has been suggested. Dogs survive well in the wild and the more moist climate of the time would have suited them. They lived in the same territories with the Tasmanian tigers, and assuming the dog was

More deeply engraved in the rock and of a rougher nature, this representation of the thylacine comes from the Burrup Peninsula near the city of Dampier, Western Australia.

large, would have competed with thylacines for food. Dogs have a very effective pack-hunting technique, and will eat a wide range of food, including dead, putrefying flesh and even vegetation if necessary, as well as being able to kill thylacines. In competing for food, the dogs have a distinct ecological advantage in that Tasmanian tigers would not eat vegetation nor dead, putrefying flesh, and are not known to employ any form of pack hunting technique.

The suggestion has been made that the dingo found its way to Australia by sea, and that this occurred sometime after 8,000 years ago, despite dingoes not appearing in an archeological context until 3,000 years ago. Dingoes must have arrived after the separation of Tasmania and the other islands from the Australian continent because they have never been found in Tasmania. They would have been strong competition for the Tasmanian tigers for the same reasons as dogs, and although the dingo's origins are uncertain, it is not believed to have evolved from the dog.

The first Aboriginal arrivals on the Australian continent are thought to have occurred some 20,000 years ago, although recent discoveries in caves in south-west Tasmania have shown human life there some 50,000 years ago.

CHAPTER TWO — EARLIEST TRACES

There is absolutely no doubt that the early Australian Aborigines knew thylacine and co-existed with the species. Dr I.M. Crawford of the Western Australian Museum has provided a photograph of an Aboriginal cave painting, which clearly shows the features of a Tasmanian tiger. The cave painting is from the Kimberleys, unfortunately Dr Crawford was unable to date it to either before or after a cultural change that took place 3,000 years ago. Mrs M. Sack tells of a tape she made in 1964, when she was the only guest at a farewell party offered in her honour by tribal elders from the Marble Bar area in Western Australia. A man from Pilbara, aged about 70, sang "a song about a dingo but the leader was the one with the stripes" in reference to the leader of the dingoes. This particular tribe spent a lot of

The Aboriginal rock art in the Kimberley's also includes representations of thylacines, as the one above, which is life size (145 cm).

time in the caves of the Hamersley Ranges before the arrival of Europeans, and this could well be the only reported case of oral reference to the Tasmanian tiger in Australian Aboriginal culture.

That the legend has been perpetuated indicates that the image may have been refreshed by visits to the sacred cave paintings, or that a visit to the caves had inspired the creation of a new song in comparatively recent times. It is also possible that the song may have absolutely nothing to do with the paintings, and could be a true oral account dating back to the time when Aborigines and Tasmanian tigers co-existed, perhaps even less than three thousand years ago.

Descriptions of Aboriginal monochrome ochre paintings from sites at Death Adder Creek and Cadell River in Arnhem Land portray various animals, some of which are most certainly Tasmanian tigers, as well as dogs and dingoes. The paintings are associated with Mimi art, which, according to Aboriginal legend, were drawn by the spirits or "old people". It is significant that the dogs and dingoes were drawn in the later post-Mimi period of cultural change using the x-ray style

The thylacine was also present in the northern part of the continent and particularly in Arnhem Land from where all the following Aboriginal paintings, art and engravings originated:
1 - Thylacine with a three-pronged spear in its back, Upper East Alligator River (Arnhem Land)
2 - Obiri (Kakadu National Park)
3 - Magarni, near Nabarlek (Arnhem Land)
4 - Malgawo, on a tributary of the Mann River (Arnhem Land)
5 - Life size (154 cm) depiction, Namalawu, Upper Magala Creek (Arnhem Land)
6 - Thylacine over hunter, in the dynamic figures style of paintings, Mt Gilruth (Arnhem Land).

1

2

3

4

5

6

Another spectacular Aboriginal rock painting from the Northern Territory.

of art. There are two paintings of Tasmanian tigers at the Obiri Rocks site near Oenpilli, one of which is very lifelike. Both paintings, like those from various other sites, are from the pre-x-ray period.

When the animals became extinct in the area, the Aborigines would have ceased painting them. The different techniques used to paint the dogs and dingoes suggest that these two species did not co-exist with the Tasmanian tiger, at least not in Arnhem Land, a theory that would suggest the estimate of 8,600 years is closer to the mark for the arrival of dogs on the continent.

The continental Tasmanian tigers may well have been hunted by the Aborigines, but it is unlikely that this would be a major cause of extinction. It is more likely that competition with dogs and dingoes was the main factor contributing to the tiger's elimination from the Australian mainland.

It is difficult to accept that tigers would have lived on the Australian continent for such a long time, only to disappear in such a relatively short period.

The species was known as lagunta or corinna to the Aborigines on the east coast of Tasmania, and as loarrina to the north-west tribes. Both the southern tribes and the Bruny Island tribes had two names for the thylacine, namely laoonana and ka-nunnah. Although Tasmanian tigers have not been recorded on Bruny Island, the existence of two names suggests the islanders knew the animal well, and probably encountered it on visits to the Tasmanian mainland.

Thylacines were of some importance as a food source for the Tasmanian Aborigines, and in the diaries of George Robinson, a man who walked over most of Tasmania in the 1830 - 34 period, it is noted that Aborigines killed tigers whenever possible. The predatory activity of the Aborigines towards thylacines could not have been excessive, as the Aboriginal population of Tasmania was small, probably not exceeding 5,000 and possibly even less. It could nonetheless have helped prevent any major increase in thylacine numbers, particularly if hunting was concentrated on juvenile specimens.

Tasmanian Aborigines and tigers co-existed into modern times, and since neither dogs nor dingoes can be held responsible for the decline of the tiger in Tasmania, reasons for its disappearance must be sought elsewhere.

1- Information taken from *"Prehistoric Mammals of Western Australia"* by Ken McNamara and Peter Murray (W.A. Museum - 1985).

Chapter Three
Anatomy: a

closer look

There are only a few measurements of the Tasmanian tiger, and even fewer records of its weight. Much of the information available was collated by Professor of Zoology Heinz Moeller, who used museum specimens as well as sources from older literature. The head and body lengths range from 851 to 1,181 mm (average 1,086 mm), tail length 331 to 610 mm (average 534 mm), height at the shoulder about 560 mm, the whole weighing about 25 kg. Several old-timers made the point that Tasmanian tigers were frequently in the region of 7 feet long (2m133). In 1980 Sawley, who lived in north-west Tasmania in the 1900-1930 period, claimed that *"a tiger shot at McKay's property at Trowutta was the biggest ever known in this area measuring 9 feet (2m716) from the tip of nose to tip of tail"*. It is possible that animals larger than those ever sent to museums, universities and zoos did indeed exist.

The head is remarkably like that of a dog, although the ears are shorter and rounder. The lateral view of the skull shows the large gape of the Tasmanian tiger. Teeth have always fascinated some zoologists and those of the thylacine are no exception. In 1968 Moeller gave a very detailed description of the dentition, which is typically carnivorous in form, and well adapted for shearing meat.

The only milk dentition which has been described was a deciduous premolar tooth found by W.H. Flower in 1867. The dental formula is 4/3; 1/1; 3/4; 4/4, which means there are four incisor teeth in the upper jaw and three in the lower, one canine tooth in both the upper and lower sets of teeth, 3 upper pre-molars and 4 lower, and finally four molars in the upper and lower jaws. The formula relates to one side of the skull, hence the total number of teeth is 48 (dental formula multiplied by 2).

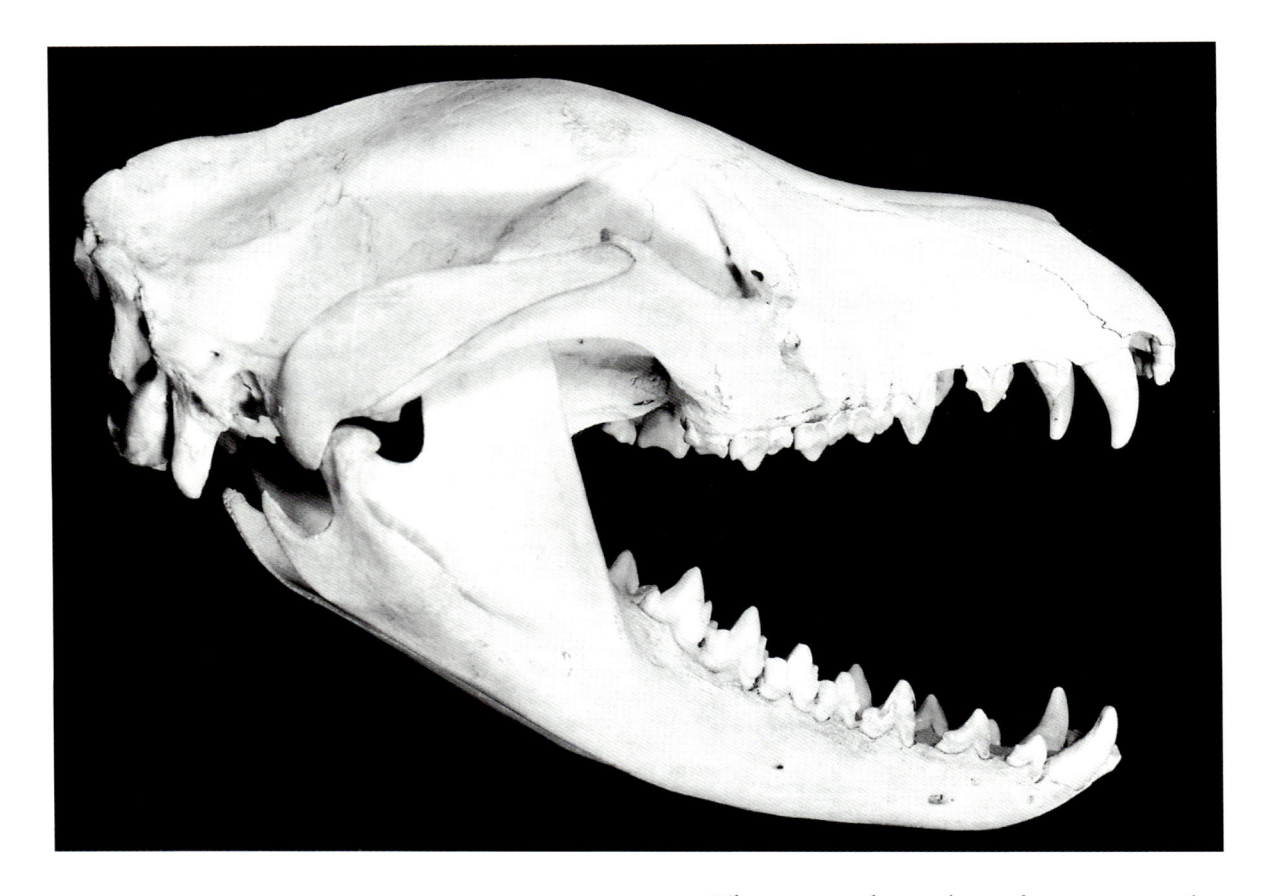

Previous page: The skeleton of the Tasmanian tiger.
Above: Lateral view of the skull.
Right: A yawning Tasmanian tiger which shows the enormous gape enabling it to seize its prey.

The unequal number of incisors in the upper and lower jaws is one of the features that differs from dogs. It was found that the teeth of the thylacine do not suffer greatly from wear, remaining sharp and capable of shearing flesh throughout life. This is in contrast to the Tasmanian devil, which suffers significant tooth wear and breakage from crushing bones.

The thylacine's brain has been described in detail, and is larger than any of the tiger's relatives, and greater than might be expected in proportion to the species' other features. On the other hand, the scent-related olfactory lobes are relatively smaller than could be expected, and this provides a clue to the thylacine's lifestyle. Being a relatively tall animal able to look over the top of grass and scrub, sight is more important than scent in seeking and hunting down prey. In 1970, Moeller noted that the neocortex of the brain is well developed. The ridges of this structure are related to intelligence - perhaps the thylacine is "brighter" than its relatives.

The 13 to 19 dark brown coloured bands extending from the posterior thoracic region to the butt of the tail are the most characteristic feature of the thylacine. These bands provide contrast to the brown, light brown or sandy colour of the body. The

CHAPTER THREE — ANATOMY: A CLOSER LOOK

forward bands extend only a short distance from the midline, whilst the longest bands occurring on the rump extend laterally as far as the upper thigh.

The posterior rump bands are short and extend on to the butt of the tail. The number of stripes varies from individual to individual. It has been noted that this pattern is rare amongst mammals, but is found in animals that live in savanna woodland or forest areas, which suggests that the thylacine is an animal of these habitats.

The body hair is short and dense, usually a fawn to yellow-brown colour, though some individuals, usually young, are of a darker shade. The ventral belly fur is a creamy colour. The tail fur is close and tight, in contrast to that of the dog, which is often longish with a brush. The female has four nipples and a backward opening pouch, which is an advantage when moving through grass, as it prevents the young from being speared by twigs and other sharp objects. Various descriptions of the male thylacine depict a pouch in which the testes are sheltered, however the photograph shows the testes clearly exposed.

The tail is very different from that of the canids. It appears as a continuation of the body, with a strong musculature at the base that extends down to the tail. This lends a rigidity to the tail, making the animal appear somewhat clumsy when turning around. Some people describe the movement as being like a ship turning into the wind.

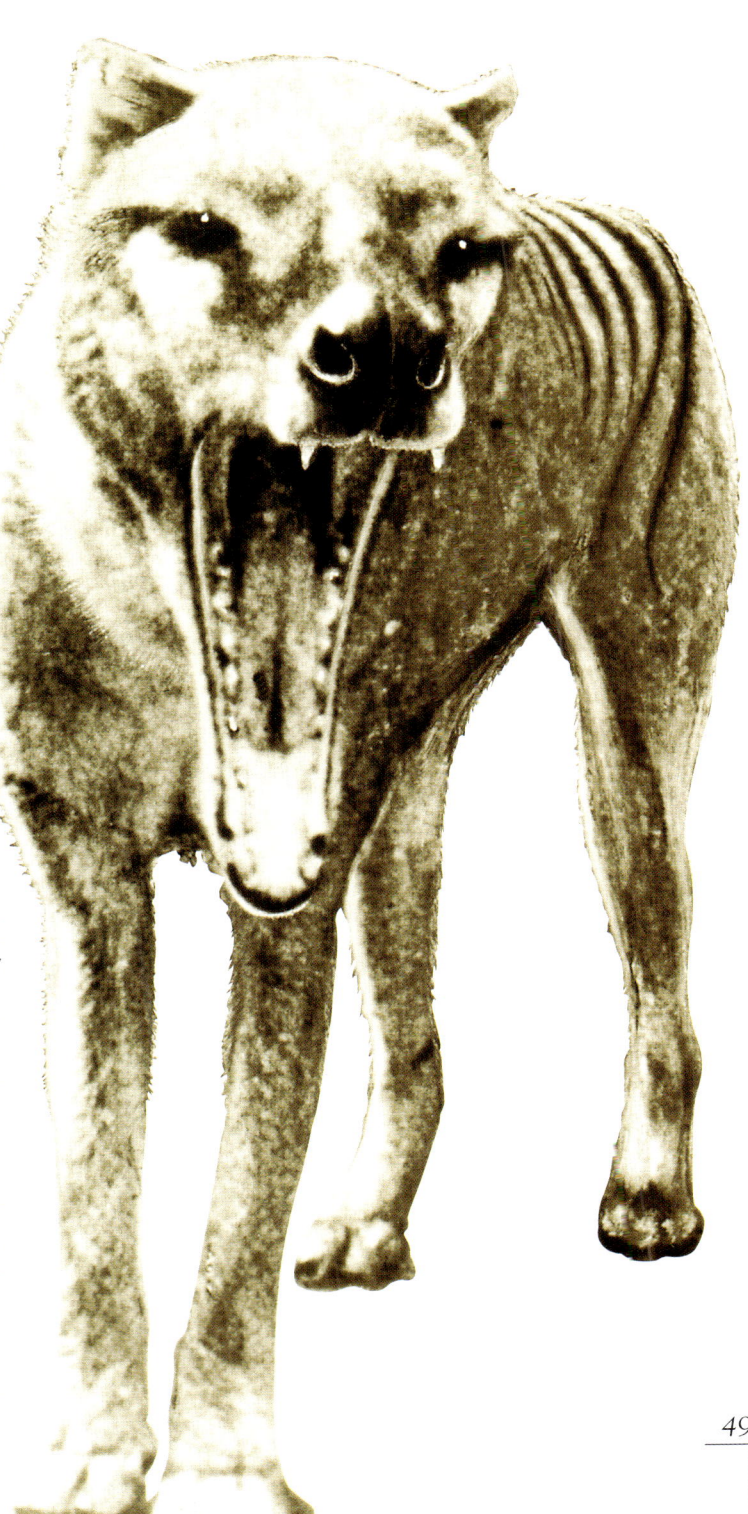

Skin of a Tasmanian tiger as it would have been pegged out to dry prior to sale to a skin merchant - how many tigers ended their days in this manner?

It has been reported that the Tasmanian tiger can leap like a kangaroo, however a very detailed study of the ankle joint shows no evidence of a hopping habit. No features in the limbs have been found which were not compatible with a quadripedal habit. It may well be that the thylacine can raise itself up on its hind legs when moving through grass in order to get a better view, which is something dogs also do.

Another peculiar feature of the skeleton is the absence of the marsupial bones in the pelvic girdle. These bones are found in all marsupials as a support for the pouch and it might be expected that an active predatory species would have these bones to support the pouch when the animal is hunting.

Thylacine footprints are of prime importance in any evaluation of field evidence to determine the presence or otherwise of the species. Unfortunately, the previously described similarities with the dog are equally evident in the feet. Thylacine feet are shown in the illustration. The plantar surface of the forefoot is separated from the toe pads by a prominent space seen in the spoor. The toes are arranged symmetrically around the rather irregularly shaped foot pad, which has two deep grooves extending forward from the rear. The toe pads are of moderate size, with the figure given by Pocock being somewhat misleading in that it shows five toes on the front foot. In actual fact, one of the toes is raised above the others and leaves little impression except in soft mud. On the front foot, toes three and four, of equal size, are placed at equal distance in advance of the foot. Toes two and five are positioned at the same level as each other, but further behind toes three and four.

The hind foot is more easily identified in the field because the space separating the digits from the main pad is larger, and there is a single prominent groove running

CHAPTER THREE — ANATOMY: A CLOSER LOOK

forward from the back of the foot. The toes are arranged in similar fashion to those of the front foot, except there are only four of them. This spacing of the toes is quite obvious in the spoor, with the claws of both feet visible but less prominent. The tarsal segment of the leg bears a granulated surface that sometimes touches the ground, but only appears in the spoor under unusual circumstances. This was the case for a spoor found in a track left by a tractor, in which the hock of the animal left an impression in the steep muddy side of the tyre marks.

The tiger's gait is a trot, the distance between impressions of any one foot being about 70-80 cm, although this will vary according to the size of the animal. A tiger supposedly seen on the west coast of Tasmania in January 1970 was described as trotting like a horse. Although no information is available with regard to the running gait, it has been stated that Tasmanian

This photograph from Launceston City Park is unique because it is the only one to show the testes of the male. Many authors have stated that the testes are contained in the pouch, however this is clearly not the case. Harris' original description described the testes as pendulous and contained in 'two folds of skin or pouch'. This arrangement is much the same as that in the Tasmanian devil where the testes are contained in similar folds of skin, the 'pseudo-pouch' of Guiler & Heddle.

TASMANIAN TIGER — A LESSON TO BE LEARNT

Drawings of Thylacinus feet based on dry specimens:
a. right hind foot
b. right forefoot
c. & d. right forefoot and hind foot in normal walking position

A lot of time has been spent walking through the bush looking for footprints. The front foot has four toes and the back has five. The pad of each foot has two grooves running forward from the rear side and extending about halfway across the pad. These grooves are important in identifying the print.

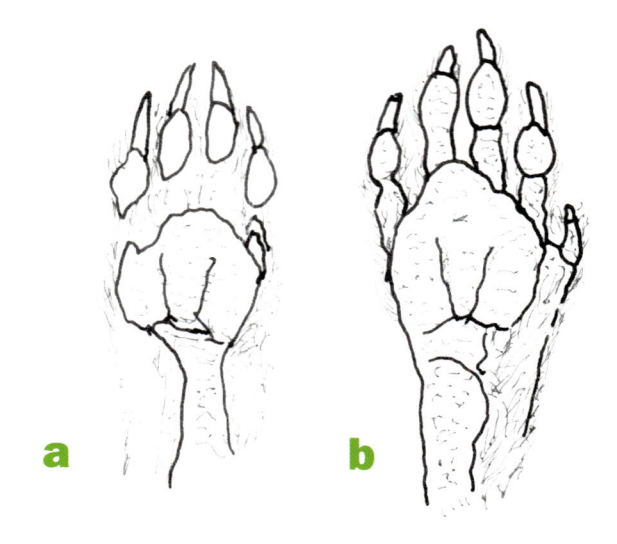

Sketches at actual size of footprints:
a. left front foot of Tasmanian tiger
b. left rear foot of Tasmanian tiger
c. dog
d. rear foot of Tasmanian devil
e. rear foot of Tasmanian devil
f. left front foot of wombat

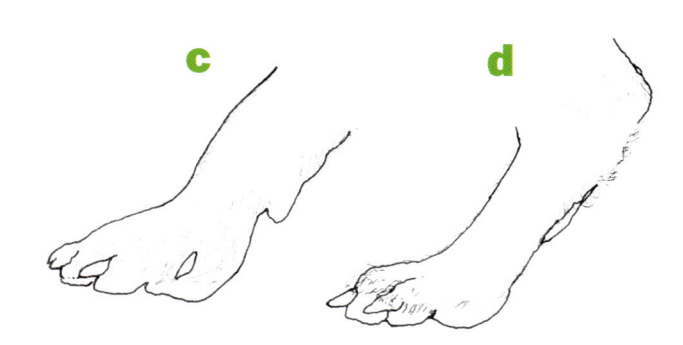

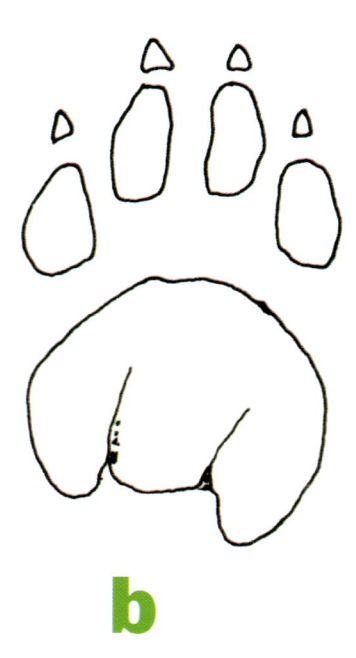

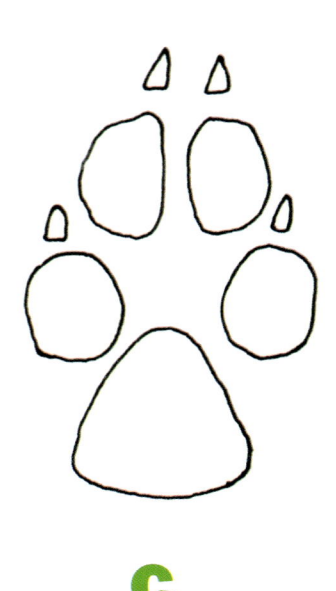

CHAPTER THREE — ANATOMY: A CLOSER LOOK

tigers broke into a shambling canter when pressed, which is similar to the running gait of the Tasmanian devil.

A tiger which had escaped from a snare was described as "bounding like a lion", and there are other references to tigers hopping like kangaroos, especially when in hard chase. Trappers have been quoted as saying that the thylacine takes several kangaroo-like hops before assuming a normal gait, which could well be the "bounding like a lion".

The bound is one of the most persistent legends to have sprung up about the Tasmanian tiger, but apart from a few bounds in the initial stages of a chase, it is very doubtful that the animal could hop any great distance. There is nothing abnormal about the hind limb and it has been shown that there was a resemblance between the convexity of both the thylacine's and the kangaroo's crosswise heel bone. This is a feature found in both running and jumping animals and cannot therefore be considered convincing evidence for bipedal locomotion. There would appear to be no unusual features in the muscles, nervous system or bone structure of the hind limb that would have enabled the tiger to indulge in bipedal frolics. Furthermore, the tiger's tail is rigid and would surely get in the way of any hopping movement.

A 1982 study (Keast) showed that the proportion of the limb segments between the tiger's ankle and toes in relation to the total limb are shorter than a wolf's, and there is no anatomical suggestion that thylacine hind limbs are in any way adapted for bounding. Like dogs in similar circumstances, Tasmanian tigers have been seen to rise up on their hind legs in tall grass, which may explain the legend.

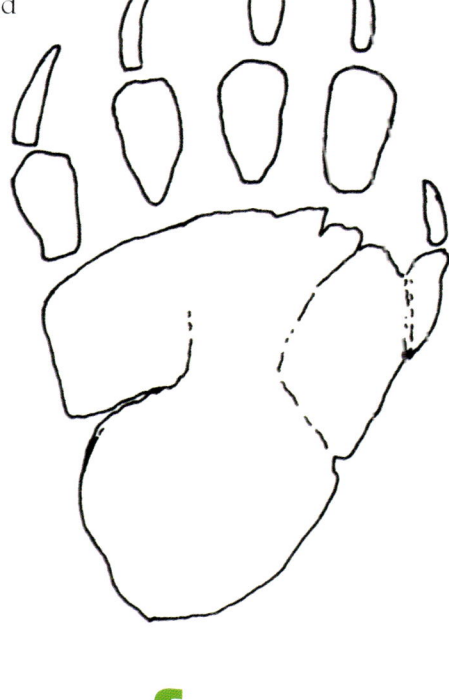

d   e   f

Reports also indicate that the Tasmanian tiger can leap like a cat, one person stating that the thylacine could jump over a six metre high wood heap from a standing start without touching the wood. A thylacine shot by Wilf Batty had been in the area for several days and had been spotted in an adjoining paddock where it had jumped a 145 cm (4 feet 9 inches) fence, only touching the top strand with its front feet. This would appear to be within the capabilities of the species, however the woodpile leap is hard to accept. The tigers are fairly agile, with an 1863 report of specimens jumping from beam to beam in their compound.

Although highly unlikely, another legend relates that the tigers are fearful of water and cannot swim. The earliest evidence to the contrary comes from Tasmanian Aborigines who said the thylacine was a strong swimmer. This maybe the origin of Swainson's notion of the tiger being an aquatic species that caught fish. Other observations of Tasmanian tigers swimming are reported, however a suggestion that tigers frequented the seashore, eating carrion and probably swimming after fish is a little far-fetched. This could well be an extension of claims that the tigers were found in rocky coastal areas, eating crabs and taking refuge in bays and caves.

The Tasmanian tiger was the largest marsupial predator to live in modern times. The other marsupial carnivores are either medium sized like the Tasmanian devil, small like the tiger cat and native cats, or tiny like the marsupial mice. In 1968, Moeller gave a very detailed comparison of Thylacinus anatomy to that of other Dasyure marsupials, showing that the body proportions were similar to Dasyurus but different from the tiger cat - Dasyurops - and

Tasmanian tiger, Beaumaris Zoo.
Below: This is the stance that would be used when looking over high vegetation. Both the tail and the feet are in a typical kangaroo posture.
Right: Note the position of the hind foot on the ground.
(drawings inspired from photographs)

Sarcophilus, in that the rear limbs are longer. Plotting thylacine skull length on a logarithmic scale against the length of the fore and rear limbs gives a similar straight-line relationship and hence similar proportions to those of the native and tiger cats. The curve for the Tasmanian devil is different.

Over and above their academic value, these conclusions reflect the life form of the species. The native cat is an active predator and the tiger cat, although not as active, still leads a predatory existence requiring a considerable degree of agility. On the other hand, the Tasmanian devil is a slow-moving scavenger with little need for agility, which is evident in the relatively poor development of the rear limbs. Being an active predator, the thylacine requires considerable power in its back legs, and this is confirmed by the limb proportions.

Moeller also compares the anatomy and body proportions of the Tasmanian tiger with a selection of canine runners, namely Canis the Indian wild dog, dhole (Cuon), the striped hyena, and the maned wolf Chrysocyon. He found that the length of the thylacine's forelimb is shorter than the others except for the dhole, which also resembles the thylacine with its short neck. The tiger's tail is rigid and much longer than any of the canines, a long tail being a common feature of marsupials. Moeller also points out the feline runners, in particular the serval and the cheetah, have similar body proportions to the canines, concluding that in this respect, the Tasmanian tiger more closely resembles a leopard than any canine species.

The thylacine and the leopard both have short legs and long tails, but the leopard has a shorter skull. The morphological similarities are intriguing and quite fortuitous, their habits are not the same. The notion of carrying a large piece of meat up a tree may be attractive, but remains unsupported by even the most imaginative of tiger tales. Moeller's study also shows that the thylacine is not as fully adapted to a runner's life as the canine species are, and his main conclusion settles on no real relationship between Tasmanian tiger and wolf body proportions, any resemblance being superficial, although there are similarities in the shape of the foot pads.

Apart from the basic differences between marsupial skulls and those of eutherians, the third and by far the largest group of true placental mammals, Moeller shows other differences with regard to the nasal bones and arches, with thylacine males having wider foreheads and jugals than their female counterparts, differences upon which Krefft established his new species. The tiger's forehead is wider than a wolf's, whilst jugular widths are similar. From cranium to midpoint, a large wolf's skull is greater than a large tiger's, but the major difference between the skulls of the two species is the width at the penultimate molar, 8 mm greater in wolf skulls, more in larger animals.

Above and below:
Dorsal and ventral views of the skull of a dog.
Right page:
Dorsal and ventral views of the skull of a Tasmanian tiger.

The brain case is 14 mm longer in wolves with a capacity of 100 ml, far superior to the thylacine's 60 ml. Brain volume to body length curves slope differently, indicating little brain growth in older tigers. The dental formulae are 4/3 - 1/1 - 3/4 - 4/4 for the thylacine and 3/3 - 1/1 - 4/4 - 2/3 for the wolf, with the thylacine having weaker carnassial teeth, perhaps associated with the smaller size of prey available to them.

Comparison to wolves was pursued by Keast (1982), who also compared the body proportions of the tiger cat, marsupial mouse, weasel and marten. Amongst other things, Keast found that thylacines had shorter metatarsal and metacarpal segments in relation to total limb length, concluding that, despite superficial resemblance, the only similarity in body proportion was increased neck lengths for wolves and Tasmanian tigers in comparison to other Dasyures. Keast also noted that although Tasmanian tigers are capable predators, they are not as specialised as wolves in this role.

We can therefore assume that the running and hunting habits of the wolf and the Tasmanian tiger have led to superficial resemblance, despite quite different body proportions. Skulls are the most similar elements, but the wolf

CHAPTER THREE — ANATOMY: A CLOSER LOOK

has a much larger brain. The overall build gives the wolf an agile aspect, whilst thylacine appears slower and more clumsy, with an ungainly, heavy tail.

Marsupial vocalisations do not reach the high degree of development found in eutherian mammals, and apart from encounters in combat situations, marsupials tend to be quiet animals. When watching marsupials, it becomes apparent they make a loud noise that can only be accomplished as a result of considerable effort. The production of sound does not appear easy, and it is not known how many sounds the Tasmanian tiger can actually make, nor the circumstances in which they do. Le Souef recorded a coughing sound, and trappers have described the same sort of noise, although it remains unclear what the sound means.

A distinct yapping sound likened to that of a dog barking was apparently used when hunting. According to a variety of sources, the yap-yap or nasal yaff-yaff noise was high-pitched, with the second yap in a lower pitch, following the first yap very quickly, almost like an echo. When irritated, the thylacine uttered a low growl, the inspiration being rapid and accompanied by a harsh hissing noise issued as a warning when the animal was excited.

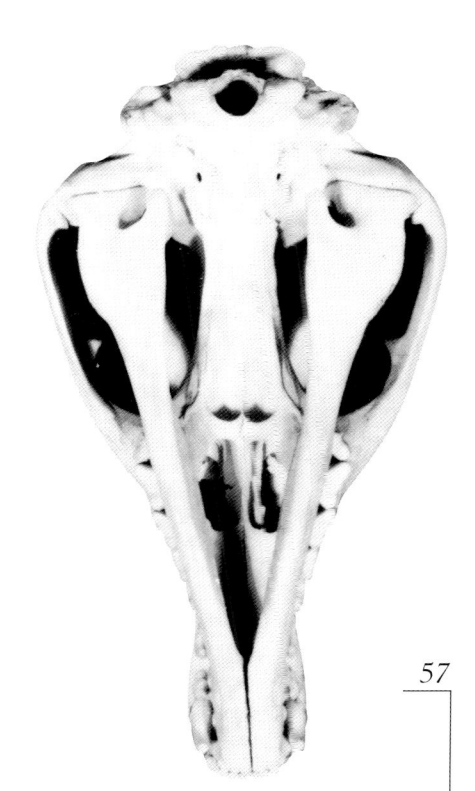

TASMANIAN TIGER — A LESSON TO BE LEARNT

Chapter Four
Thylacine's

Tasmania: last home

A little known part of the world, the island of Tasmania lies off the south-eastern end of the Australian continent, separated from the mainland by Bass Strait. Once a far-flung part of the British Empire, it is now a State of the Commonwealth of Australia.

It is generally agreed that the island was formerly connected to the continent by a land bridge, and this enabled animals to move freely between what is now Tasmania and the Australian mainland. Rising sea levels submerged the land bridge, but there is some doubt as to when this actually occurred. The period about 10,000 to 13,000 years ago is frequently accepted, but it may have been as long as 140,000 years before present times.

The separation isolated the Tasmanian fauna, which now differs in some respects to that of Southern Australia. The Tasmanian tiger and the Tasmanian devil, both now extinct on continental Australia, survived (at least until modern times in the case of the thylacine) on the island-State. It is not known why several species of wallaby, found today in southern Victoria, did not cross the bridge to Tasmania before it was cut off from the mainland.

Tasmania presents an entirely different habitat from the one encountered by thylacines that once lived on continental Australia. With the exception of the Great Dividing Range, the continent, ranging from savannah grasslands to desert, is much drier than Tasmania. The mainland plains are vast, reaching as far as the eye can see, even from planes flying at high altitudes. The rainfall over these enormous areas varies greatly from the tropical, monsoonal conditions of Arnhem Land and the Northern Territory, to the restricted rainfall in the centre of the continent.

The temperatures in mainland Australia are much higher, with over 40°C being commonplace in summer throughout the plains country. Such temperatures are only encountered on a few occasions in Tasmania in any given year, and water is much less readily available on the continent.

Previous page: Belvoir Plains (west coast), excellent potential habitat for Tasmania tigers. Cradle Mountain National Park is in the background. The plains and scrub offer shelter as well as hunting opportunities. Large numbers of kangaroos and wallabies live in these forests. *(map: 1)*

Below: Rainforest, Mt. Anne, S.W. Tasmania. Old, over-mature forest with Richea plants, some of which are over two metres tall, and tangled fallen logs, all encrusted in thick moss. Movement through this country is difficult. There is no grazing for species upon which thylacines would prey, hence they did not live in this type of country. *(map: 2)*

Despite these climatic differences, the Tasmanian tiger survived on the Australian mainland for many centuries, and its disappearance cannot be linked to climatic changes, as the climate has differed little in the last 3,000 or more years. It may be that thylacine was already on its way to extinction before man's interference in Tasmania.

The island's 68,000 square kilometre surface area is approximately the same as West Virginia in the United States and slightly smaller than Ireland, and is noted for its variety of scenery. Much of Tasmania is mountainous, with the Ben Lomond

CHAPTER FOUR — TASMANIA: THYLACINE'S LAST HOME

A winter mountain scene, Barn Bluff and Lake Will, in the Cradle Mountain National Park. The rock and scree slopes are snowy, with the forest and buttongrass plains below but lightly covered. The snow does not last long enough to force animals to migrate to lower ground. Thylacines would have hunted around the edges of the forest. (map: 3)

Next spread: Cox's Bight, South Coast, from the New Harbour region. The mountains descend down to the sea on the south coast, leaving little coastal dune habitat, except in bays where valleys and creeks meet the sea. The buttongrass plains behind the beaches offer little food for grazing animal. *(map: 4)*

massif rising to about 1,700 m in the north-east, continuing southward as a range of hills, the Eastern Tiers, which reach to the south-eastern end of the island.

The wide Midlands Plain separates Ben Lomond from the other two mountain masses of Tasmania. This plain consists of open grasslands, with scattered trees and watercourses meandering across them. There is reason to believe that the plains were created by the Aborigines, who persistently fired the woodlands to create hunting grounds upon which they would pursue grazing wallabies. Some relics of the old forests are still preserved.

To the west of the Midland Plains is another mountainous region, the Central Plateau, which, as the name implies, is in the centre of Tasmania. The plateau rises steeply some 600 m to 700 m from the plains, and in some places forms cliffs, where it is known as the Western Tiers.

At its western end, the Plateau continues to become the West Coast mountains, covering most of the western part of the State. Some of the peaks in the Central Plateau and West Coast mountains are high, reaching 1,700 m, although none exceed 2,000 m.

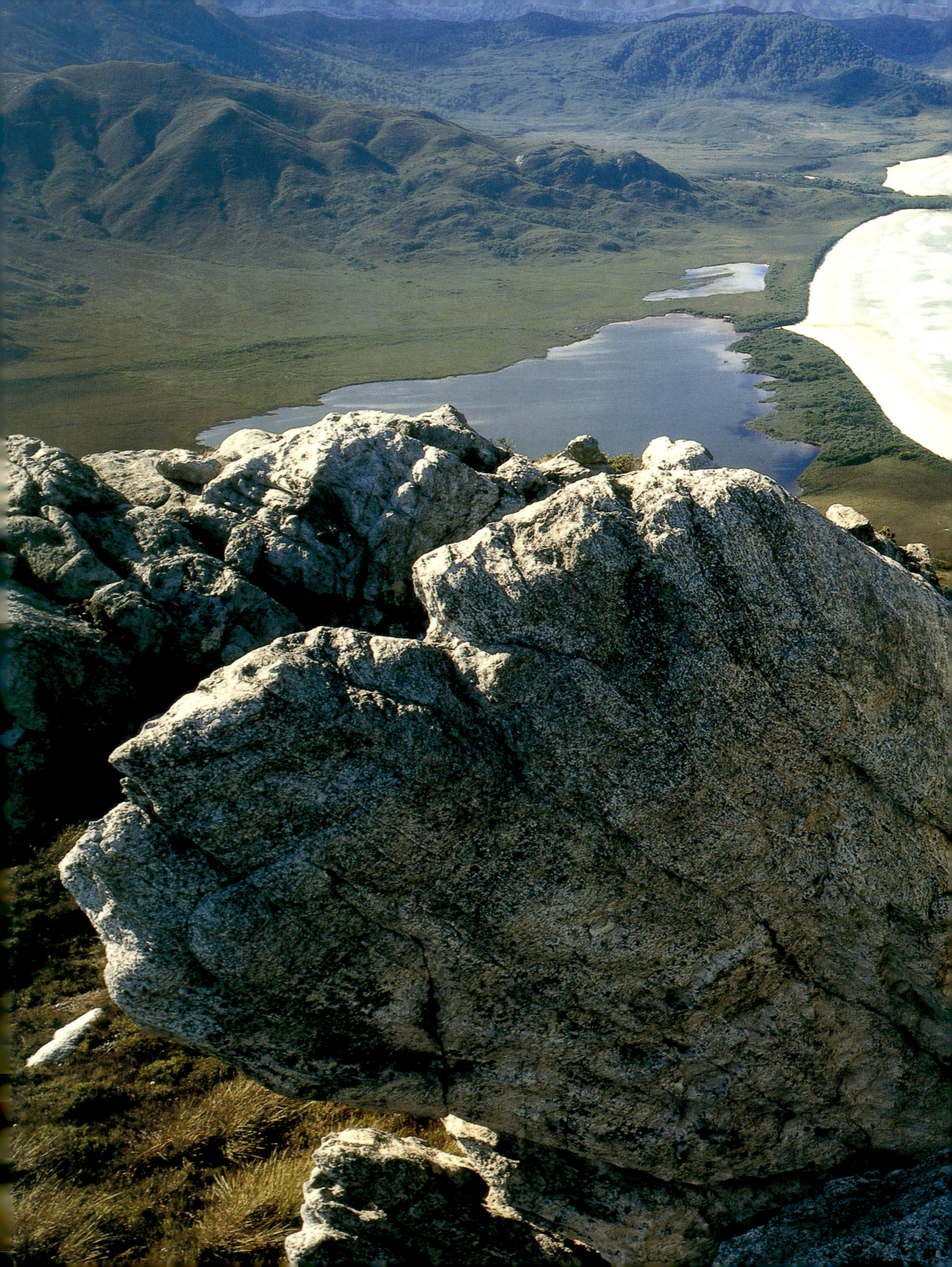

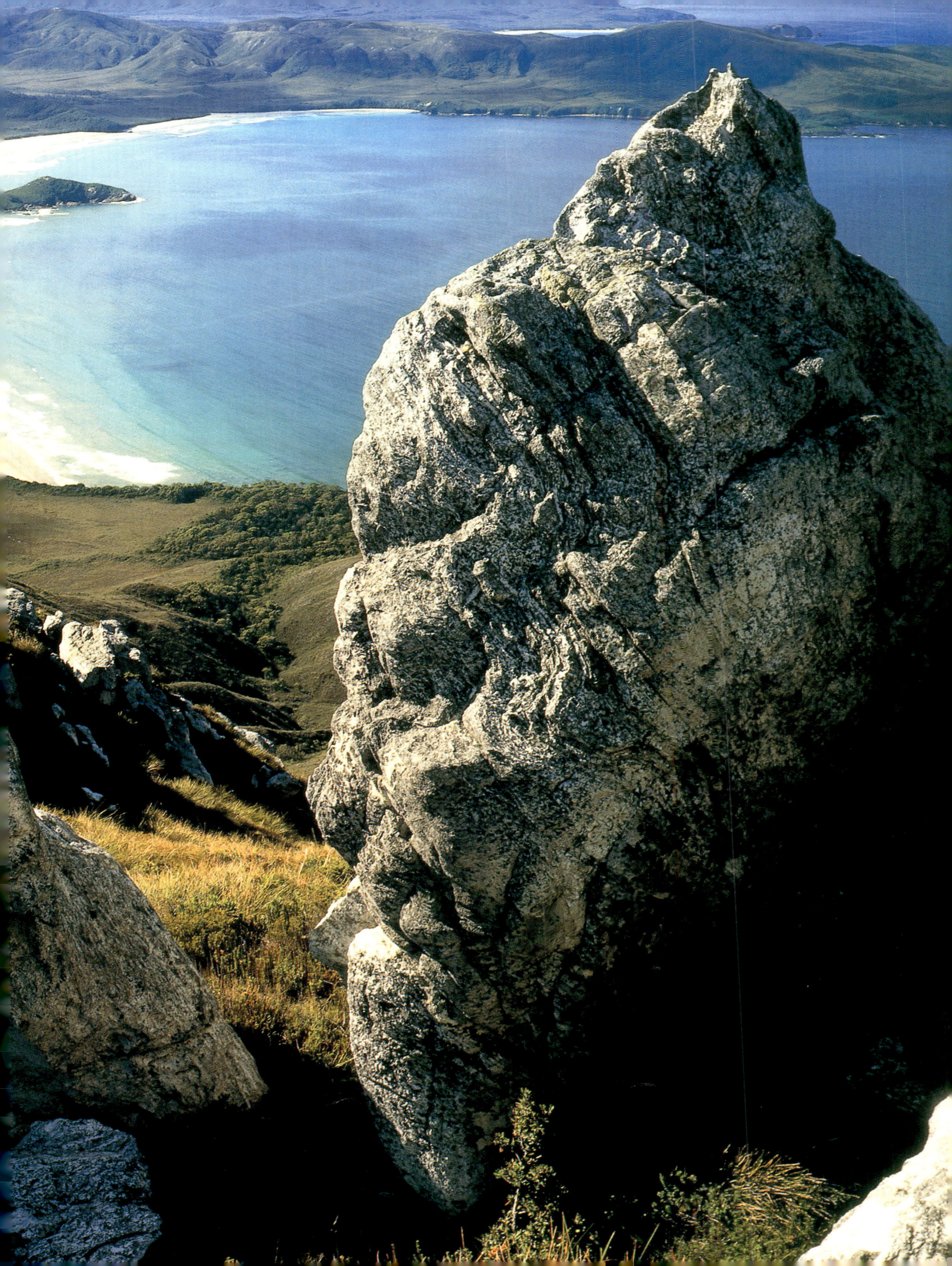

Much of Tasmania is over 1,000 metres above sea level. The height of the ranges is not the significant attribute of the landscape, the vegetation and rough terrain are the more striking features. The tops of the mountains usually display bare or lichen-covered rocks and boulders, with low, ground-hugging shrubs and occasional stunted trees in sheltered parts. Delightful to look at and very tempting to walk on, cushion plants are found in moist spots between rocks.

Alpine shrubs and plants occur at slightly lower levels forming an austral-montane moorland which offers a springtime vista of flowers. In places the cushion plant shows a bright green carpet in contrast to the rather dull greens of the other plants. All of these alpine areas are often waterlogged, with frequent little rills running into the main streams. Boggy patches are common, especially on the walking tracks.

Moving down from the mountain tops, scree slopes are prevalent. Trees, scrub and ground dwelling plants often form a dense mass in the deep gullies, which is difficult to penetrate even for the larger animals. A small stream is commonplace at the bottom of a gully, then further down the mountain the first forests of giant eucalypts, southern myrtles or blackwoods appear, forming the high canopy, with an understorey of other species of smaller trees. This is the habitat which has suffered the most from human interference. It has been burnt, grazed, cut down, ploughed, farmed, fenced, fertilised and poisoned.

Below: **Typical Central Plateau pastures, offering good hunting. Thylacines were killed in this area.** *(map: 5)*
Right: **Tasmanian highland distribution.**

The forest floor is relatively devoid of plants in some places, where the only covering is of ferns and mosses, whilst in other parts there is dense undercover of small trees and large shrubs, which make for hard going on foot. The forest and mountains are intersected by streams, but these watercourses do not constitute a serious barrier to progress unless there has been heavy rain, after which even the more gentle streams are transformed into torrents. Rainfall is very high in much of the western parts of the state, as much as 3,000 mm in places, which means the forests are almost permanently wet, making leeches common and often abundant. This type of forest offers stern opposition to any attempts at penetration, and is a habitat in which only native rats are common.

CHAPTER FOUR — TASMANIA: THYLACINE'S LAST HOME

TASMANIAN HIGHLAND DISTRIBUTION

> 1200 m
1000 - 1200 m
500 - 1000 m
0 - 500 m

❶: Picture reference

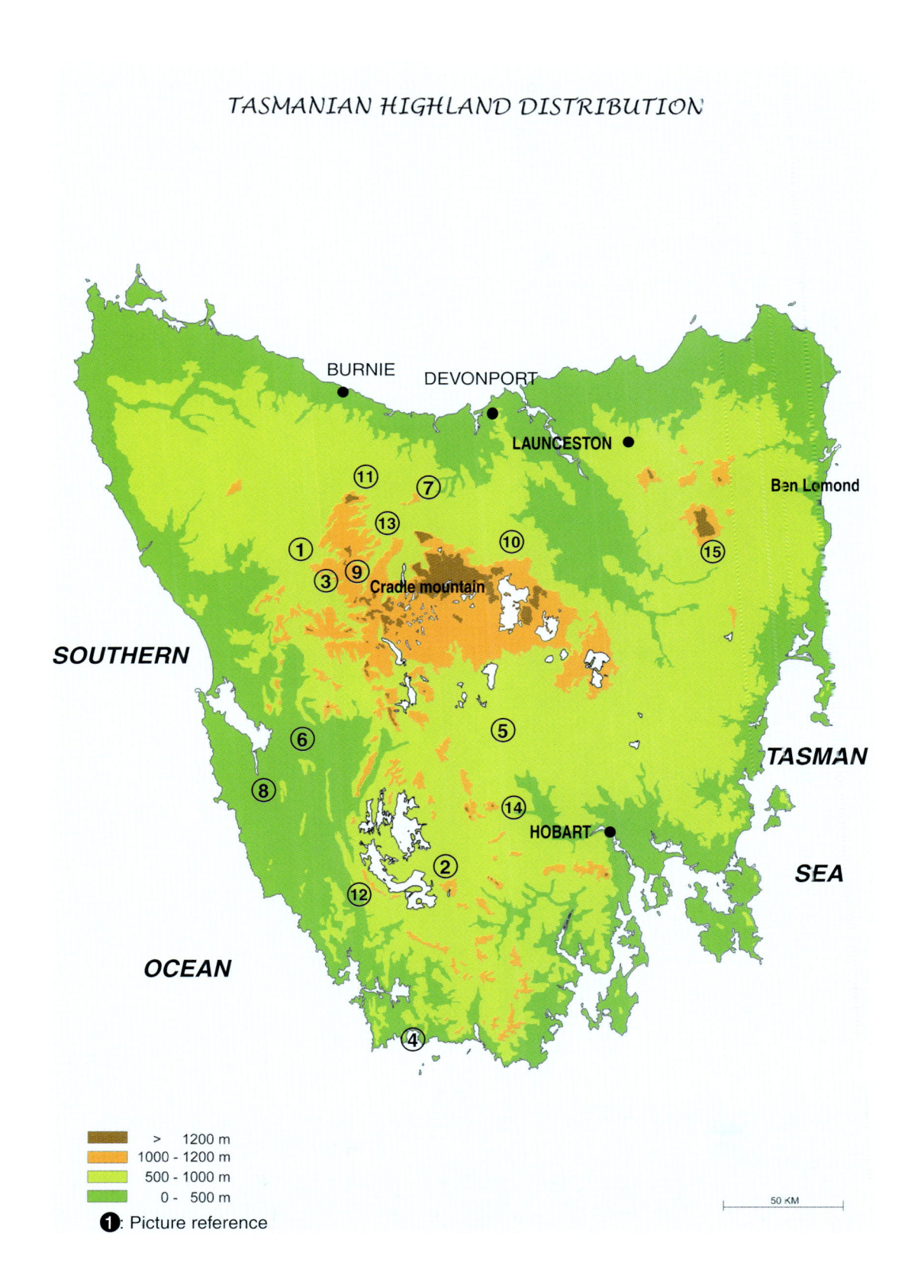

65

TASMANIAN TIGER — A LESSON TO BE LEARNT

CHAPTER FOUR — TASMANIA: THYLACINE'S LAST HOME

The forests would have been much less effective barriers if the marsupials had evolved to the size of elephants with the ability to barge through the vegetation, making tracks for other animals and providing the light necessary for grass to grow and hence provide food for herbivores to live on. The mountains, rainforest and sedgelands tend to restrict the movement of mammals, but brush possum are numerous and widespread in Tasmania, except for some parts of the south-west where they are cut off by inhospitable country.

Over most of the state, the climate is moderate, with no temperature extremes and annual rainfall of about 600 mm in many areas. However, cold weather and frequent winter snowfalls are common in the mountains and central plateau, with heavy rain at any time of the year. Snow occurs frequently above 600 m in winter, but generally does not fall for long, remaining for only a few days after a storm. At altitudes above 1,000 m, snow can remain for many days, even months. It can also be found in small drifts in early

Left: A marsupial lawn or meadow in the Upper Franklin Valley. These lawns are grazed by numerous wallabies, and wombats browse in the bushes. The plants in the foreground are *Bellendena montana* (Mountain Rocket). The forested cover around the lawns enables predators such as the thylacine to approach their prey undetected. *(map: 6)*

Above: Farming Country, Mt Roland in the background. This country was once heavily forested, but would now be of little attraction to a thylacine. *(map: 7)*

Next spread: Buttongrass heathland looking towards Eldon Bluff on the west coast. These extensive plains are typical of much of the western parts of Tasmania. The forests round the edges of these plains provided food and shelter for grazing species and thylacines would have hunted around the edges as well as on the plains, which are usually waterlogged all year round. *(map: 8)*

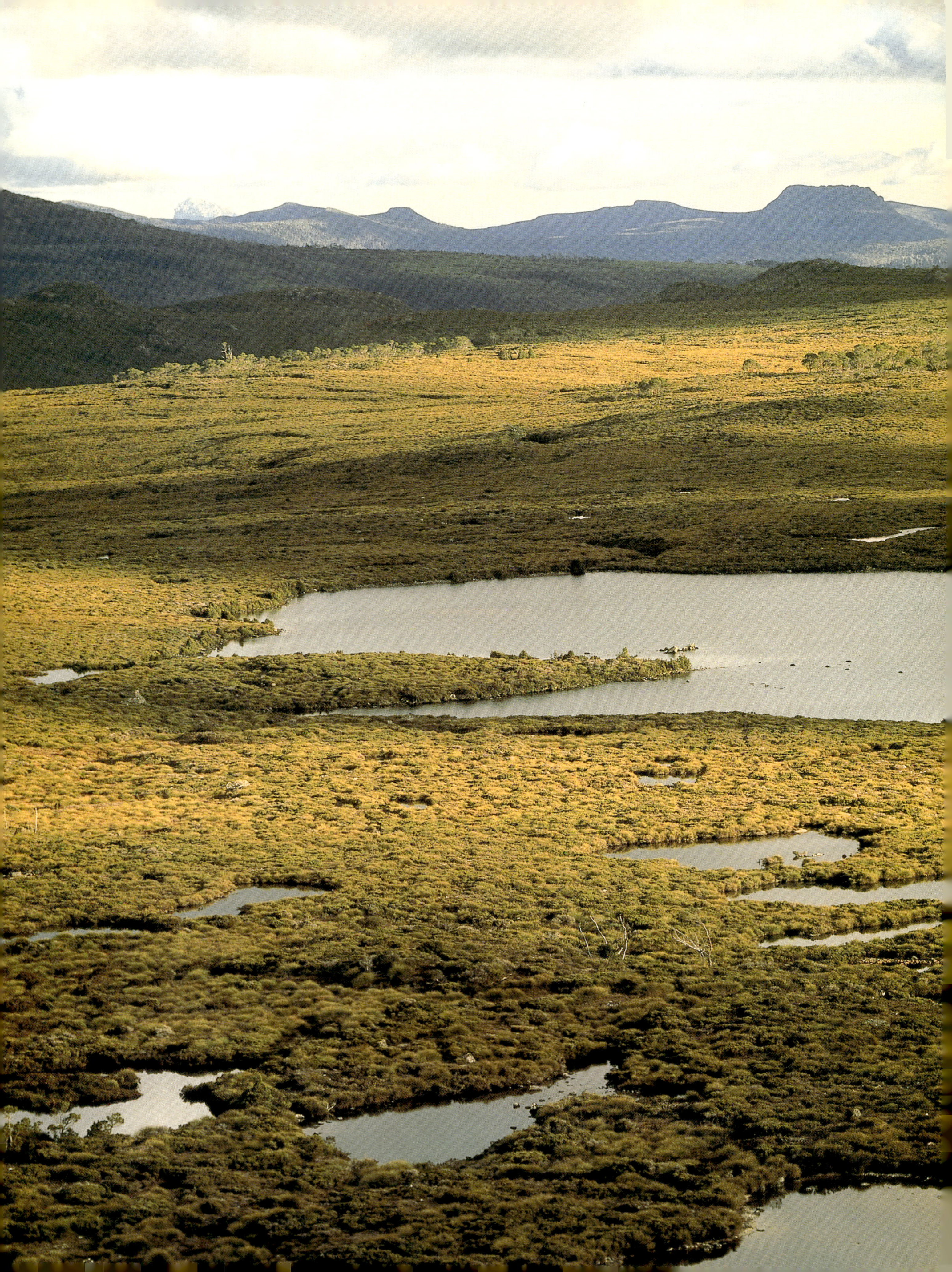

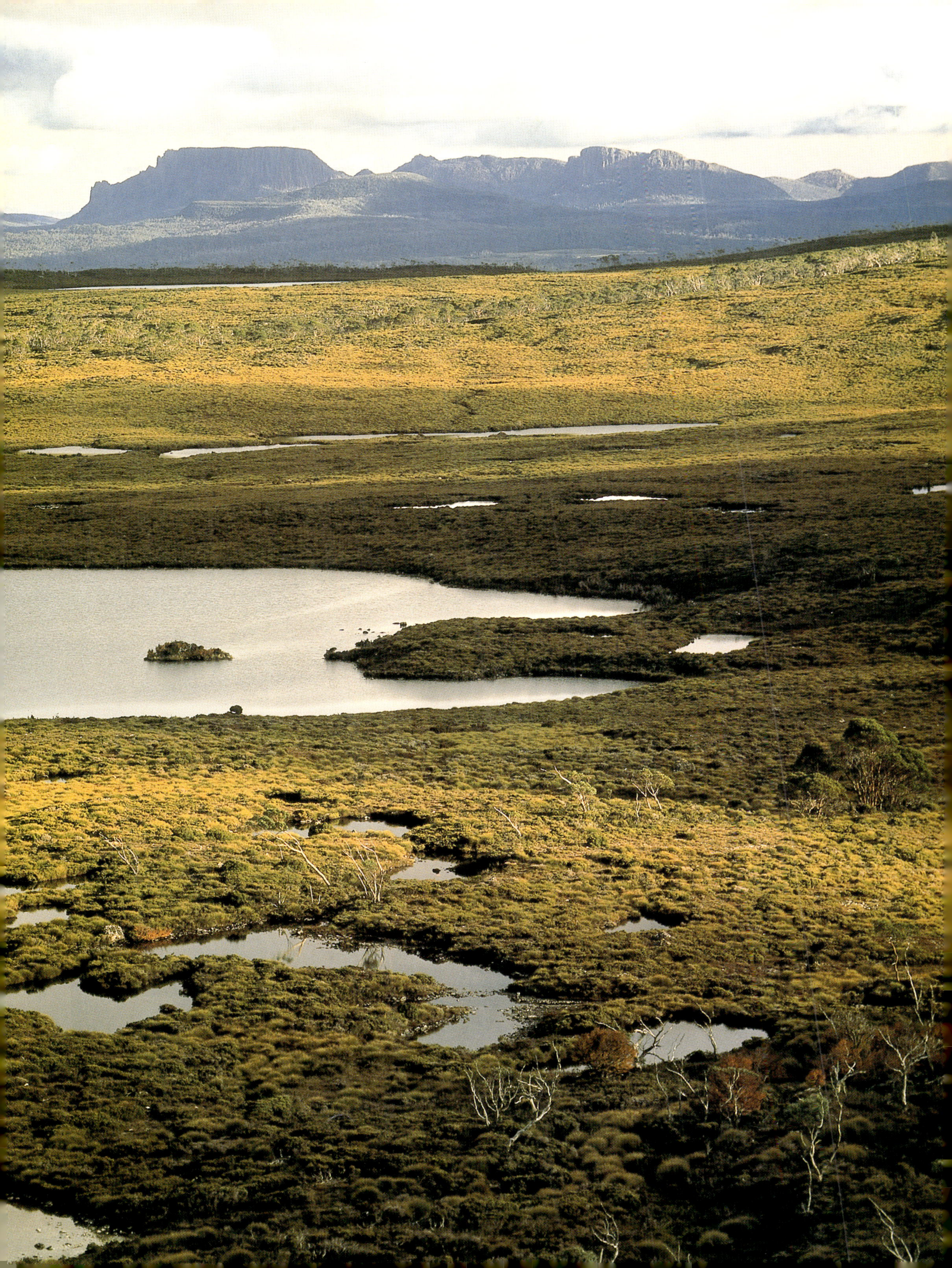

summer on the sun-sheltered side of peaks above about 1,300 m. Sporadic snowfalls can occur at altitude at any time of the year.

The Highlands are characterised by thousands of lakes, some large, but most small and un-named. Water deprivation does not feature as a limiting factor to the mammals of Tasmania, except during periods of drought on the east coast, where there are no large waterways. The major rivers drain to either the west coast, south-east coast or to the north. The west coast rivers run through deep gorges, with cliffs covered by dense rain forests on the slopes, which are substantial barriers to movement. Rapids and waterfalls are common.

Large plains are not common on the island, although much of the Tasmanian Midlands is open grassland with scattered trees. Coastal strips consist of sandhills, consolidated dunes with thick, low scrub and frequent grassy, marshy areas immediately behind the dunes. Areas where rainfall is high have sedgeland plains where the soil is highly acid with no trees, except along watercourses. These lands offer little shelter, with very restricted grazing and available food for large mammals. The coastline ranges from 200 m high cliffs to low-lying golden sand beaches. The east coast is very dry with only about 50 mm annual

Cradle Mountain and Dove Lake. Typical mountainous country with stunted forest, scree slopes and bare mountain tops. *(map: 9)*

CHAPTER FOUR — TASMANIA: THYLACINE'S LAST HOME

rainfall in a good year. Here the more open drier forests allow relatively easy passage for both humans and animals.

Although the Tasmanian tiger has been reported in all parts of the island, it is not known whether it lived in the more rugged and wet areas or whether it was simply sighted while passing through. It is difficult to accept that the tiger would live exclusively in the highlands or thick rainforest, but they undoubtedly inhabited the fringes of these areas, as well as in the thick scrub around the creeks in the sedgelands. The preferred habitat was open forest and woodlands, the type of country for which their colour pattern is most suited, and that provides plenty of cover for hiding, hunting and rearing young. Food is abundant in such parts, making for a favourite habitat for most of the other mammals as well.

The first humans to settle in Tasmania were Aborigines, who arrived at least 40,000 years ago, possibly crossing the land bridge. These people had a different culture to the inhabitants of the mainland, living in tribes, with a hunter-gatherer economy. They wandered their tribal lands, particularly the coastal regions, plains and rivers of the open forest, where game and other food was abundant.

Forests such as this covered much of Tasmania, and most of the species of mammals living in the State are found here. In days gone by, thylacines would have hunted here. The dead trees are the result of bushfire, and the forest is regenerating. (map: 10)

71

The Tasmanian Aborigines would have encountered thylacines frequently, both obviously favouring areas where game was plentiful. The Aborigines would have known much about the tiger, but very little of this knowledge has been recorded or passed down to us. George Robinson wrote that the Aborigines ate young tigers, which would have been easy to catch when left unattended by a mother out hunting.

The first Europeans arrived in 1803, and immediately began exploring the country. They often followed Aboriginal trails, setting up farms, and occupying traditional Aboriginal hunting lands, which brought about conflict between the new settlers and the former owners. Reasonably accurate maps were available by the 1850's, although much of the land remained unsettled.

Farms and villages sprang up on the plains and in open forest country, but much of the south-west remained "terra incognita", still unsettled to this day, and likely to remain so as it is now part of a World Heritage Wilderness.

Left below: **Manferns in a gully. Reports speak of thylacines seeking shelter in these places.** *(map: 12)*

Right below: **The Leven canyon in the north-western part of the island. Steep sides and thick forests make penetration difficult. The only ground dwelling mammals found here are small in size.** *(map: 11)*

Opposite right: **Lake Cethana near the Wilmot power station. Many Tasmanian devils and other small marsupials live in this area. Thylacines were also present in significant numbers in this region.** *(map: 13)*

CHAPTER FOUR — TASMANIA: THYLACINE'S LAST HOME

TASMANIAN TIGER — A LESSON TO BE LEARNT

CHAPTER FOUR — TASMANIA: THYLACINE'S LAST HOME

Human populations are concentrated in Hobart, Launceston and along the northern coast, with a total of about 470,000 inhabitants. Most of the south-west is uninhabited, with the central plateau population rather low for most of the year, except for restricted areas in summer, and during the hunting or fishing seasons. There are fewer permanent residents in many parts of the highlands than was the case fifty years ago.

Left: **Russell Falls at the eastern entrance of the Mt Field National Park. Certainly the most spectacular falls on the entire island, but this hilly area with thick vegetation was anything but favourable ground for the "tiger"** *(map: 14)*
Above: **Open forest, Rossarden. This type of forest is widespread in Tasmania, with many animal trails running through them.** *(map: 15)*

Conditions changed dramatically for the Tasmanian tiger after the Europeans arrived. Settlers established grazing lands and introduced new farm animals, which meant the tiger became a pest, rightly or wrongly blamed for the heavy stock losses encountered by the new farmers, and was mercilessly hunted for almost a century.

Even today, Tasmania offers habitats in many areas that would suit most, if not all, the requirements necessary for tiger survival. Food is abundant in much of the woodlands, central plateau and coastal plains, with large secluded areas providing the security and tranquillity needed for breeding. Many thousands of hectares remain undisturbed, particularly in the coastal part of the Heritage area, places where Tasmanian tigers could spend most of their lives without human interference.

TASMANIAN TIGER — A LESSON TO BE LEARNT

Chapter Five
The first eye

witnesses

Early explorers may well have been the first Europeans to come into contact with the Tasmanian tiger. The logs from Pedro de Mendonca's expedition in 1521 have been lost, and it is not known whether he landed during his epic voyage down the east coast of Australia. Abel Tasman reported strange footprints when he landed near Dunalley in 1642. He described them as similar to a feline tiger's prints, however there is no evidence at all that these prints were left by a Tasmanian tiger. Furthermore, there is no resemblance between feline tiger prints and those of the Tasmanian namesake. Tasman probably saw wombat footprints.

At Recherche Bay in May 1792, the Bruni d'Entrecasteaux expedition encountered an unusual animal that was described as "*un gros chien*" (a big dog). "*Gros*" in spoken French is used in preference to "*grand*" when talking about a dog. There is no doubt that thylacine morphology resembles a dog far more than the Tasmanian devil does, the devil being podgy and short. It can therefore be assumed that the famous French explorer was indeed referring to a Tasmanian tiger.

The first report of a thylacine sighting appeared in 1805, when Lieutenant Governor Paterson's dogs killed an animal "*of carnivorous and voracious tribe*" near Yorktown on the Tamar River. The 1810 description by Oxley leaves no doubt that it was a Tasmanian tiger, and the interesting comment was made that the stomach was capable of quick digestion. There is no doubt that the Tasmanian tiger had met up with its final tormentor by 1805 at the latest. Among the first settlers in Hobart was the official Padre to the Colony, Rev. R. Knopwood, whose diary is an account of the infant Colony's trials, events and successes. He wrote:

Previous page:

Antoine de Bruni, "chevalier d'Entrecasteaux", was commander of French King Louis XVI's naval armies division, when the king had him take charge of an expedition in search of his compatriot Lapérouse, of whom there had been no news since he called at Botany Bay in the beginning of the year 1788. Leaving France in September 1791, the expedition's two vessels, named *la Recherche* and *l'Espérance*, completed a round trip of the Australian continent, landing twice in the south of Van Diemen's Land, in April 1792 and January 1793. During these lengthy stopovers, the French naturalists collected plenty of specimens of local flora and fauna. Did they see the tiger on any of their brief and somewhat timid excursions inland? This is not out of the question, given a certain reference to a "big dog".

Below:

Description of the Tasmanian Tiger - Oxley 1810.

"*18 June, 1805. Am engaged all the morn. upon business examining the five prisoners that went into the bush. They informed me that on 2 May when they were in the wood they see (sic) a large tyger (sic) that the dog they had with them went nearly up to it and when the tyger see the men which were about 100 yards away from it, it went away. I make no doubt but here are many wild animals which we have not yet seen.*"

There is something tragic about these five unknown convicts who stole away into the alien bush and may have been the first newcomers to see a Tasmanian tiger, only to return to captivity and no doubt a good flogging for their adventures. Knopwood's report is certainly the first in which the term tiger is used to describe the thylacine, Paterson had earlier described the animal he saw as a hyena. A third reported sighting occurred in June 1805 on the banks of the Huon River.

REPORT ON SETTLEMENT

Quadrupeds and Birds. - The Quadrupeds and Feathered Tribes are the same as at the Derwent; both are free from that destructive animal to Sheep, the Native Dog, the dread of the Stock Holders in New South Wales. The only Animal unknown on the Continent is the Hyæna Opossum, but even here they are rarely seen, and I do not believe they have ever been seen at the Derwent. An Accurate description of this Animal by Col. Paterson will be found in the Note*; it flies at the approach of Man, and has not been known to do any Mischief, though apparently well formed to be the destroyer of Weaker Animals.

*Marginal note.-"The animal when dead weighed 45lbs.: the dimensions were from the Nose to the Eye 4 1/2 Inches; from the Eye to the Ear 3 3/4 I.; The Ear Round, the Eye large and Black; from the Ear to the Shoulder 1 foot; from the Shoulder to the first stripe 7 I.; from the 1st stripe to the extent of the Body 2 feet; length of the Tail 1 ft, 8 I. Stripes across the Back 20: on the Tail 3; it had 4 Tusks or Canine Teeth, each 1 I. long; Circumference of the Head below the Ears 1 F. 6 I.; the Height of the Animal 1 F. 10 I. the Hair on the body short and Smooth, of a greyish Colour; their Stripes Black; the Hair on the Ears of a light brown Colour. The form of the Animal is that of the Hyæna, at the same time strongly reminding the observer of a low wolf dog; its stomach was found filled with Kangaroo, and from its Interior construction it must be peculiarly quick of digestion."

<u>1810</u>

Account of the settlement at Port Dalrymple by J. Oxley. Natural history.

CHAPTER FIVE — THE FIRST EYE-WITNESSES

The early settlers were busy keeping themselves alive, and did not have much time for writing. The few records that do remain indicate tigers were not numerous in the early days of the Colony, and were a matter of comment when seen. In 1810 Oxley said that the "*hyena possums*" were rarely sighted, and he believed they had not been sighted at all in Hobart. This is incorrect given the three sightings recorded in the south in 1805, although the point is to be made that the tigers were by no means a common species, even in the early days of European settlement. In 1820 it was even stated that only four thylacines had been seen since the first settlers set foot on the land, probably the Knopwood and Paterson sightings, along with two specimens sent to London by Harris, who provided the first description of the species.

One of the most unexpected and interesting accounts was related by Rose de Saulces de Freycinet, whose husband, a Commander in the French Royal Navy, was

Louis-Claude de Saulces de Freycinet was commander of the *Casuarina*, one of the vessels of Nicolas Baudin's expedition (1801-1804), that made a fundamental contribution to a better knowledge of Australia. The schooner, built from casuarina wood, left Port Jackson on 18 November 1802, making its way to Bass Strait, and anchoring in Elephants Bay on King Island. In 19 days, the geography of the Hunter Islands to the north-west of Van Diemen's Land, was completed. Subsequent to this work, the coastal geography of Van Diemen's Land was complete, along with work that the French had previously done, as much in the southern-most parts as on the east coast and in the north of this large island. Appointed to captain the *Uranie*, Louis-Claude completed a further, and most important scientific voyage, from September 1817 to February 1820, during the period known in French history as the "Restoration". Contravening Royal Navy regulations, his wife, Rose-Marie, born Pinon, secretly embarked on the vessel, leaving a fascinating account of her exciting adventure. Leaving Sydney in the very last days of the year 1819, *Uranie* headed for Cape Horn, sailing around the southern-most tip of New Zealand.

given the command of the corvette *Uranie*. His mission was to undertake a circumnavigation voyage of scientific discovery from 1817 to 1820.

The expedition left the port of Toulon in France, but was to come to a dramatic end in the Falkland Islands, where the *Uranie* was disembowelled on an underwater reef that did not figure on the maps of the day. Prior to its demise, the vessel had made a long and scientifically successful journey around the globe, including an extended call in Port Jackson, now Sydney. It was at the end of this stop-over that the episode narrated by Rose de Saulces de Freycinet occurred. In deliberate violation of the very strict Navy regulations prohibiting officers to be accompanied by their spouses and more generally to have any member of the fairer sex on board, but with her husband's full consent, Rose de Saulces de Freycinet secretly took up quarters in his cabin, from where she was to write a fascinating account of her extraordinary adventure:

Right:
One of the very first maps of all of Tasmania. It was assembled from observations made by a number of English navigators. The two main natural harbours, where the urban centres of Hobart in the south and Launceston in the north were later established, are well illustrated. Inland areas had yet to be explored and only the summits visible from the coast are shown. This explains the absence of specific descriptions of the Tasmanian tiger around the beginning of the 19th century.

"*On Christmas Day 1819, at 8.30 in the evening, Uranie ploughed its way through the Heads at the entry to Port Jackson. We bid farewell to the pilot, placing into his custody a convict who had been discovered drunk in the space between the cannons and the keel. After a quick inspection that convinced us there could be no others on board, we hastily left the coast and headed off in a southerly direction. Alas, we realised our mistake the following day, when half a dozen convicts who had been hiding in the very depths of the boat appeared. Ill at ease in the heat and gnawed by hunger, they had finally decided to reveal their presence.*

"*Initially, my husband hesitated as to what to do with them. He was not particularly enthusiastic about keeping such undesirable guests on board. They were not seamen, and the majority would be of no use to him. On the other hand, we were already far from port and the wind was too unfavourable for him to consider returning. This last consideration brought about his decision to keep them on board. Our course took us around the bottom of New Zealand and even Campbell Island, much further to the south.*

"*(...) A few days later, we felt the first of the cold. Through a veil of fog we caught sight of Campbell Island, where seal hunters have bases. An icy sea broke on the island's black rocks, and the rare trees gave it a most frightening aspect, inspiring pity for the unfortunate men who, for meagre profit, had to stay in such a horrible place.*

"*One of the English convicts we had on board claimed he had spent several months on the island, hunting seals. He assured us that there was quite good anchorage on a clay sea bed in the south-eastern part of the island. We did indeed sight a low point extending out from the eastern part,*

CHAPTER FIVE — THE FIRST EYE-WITNESSES

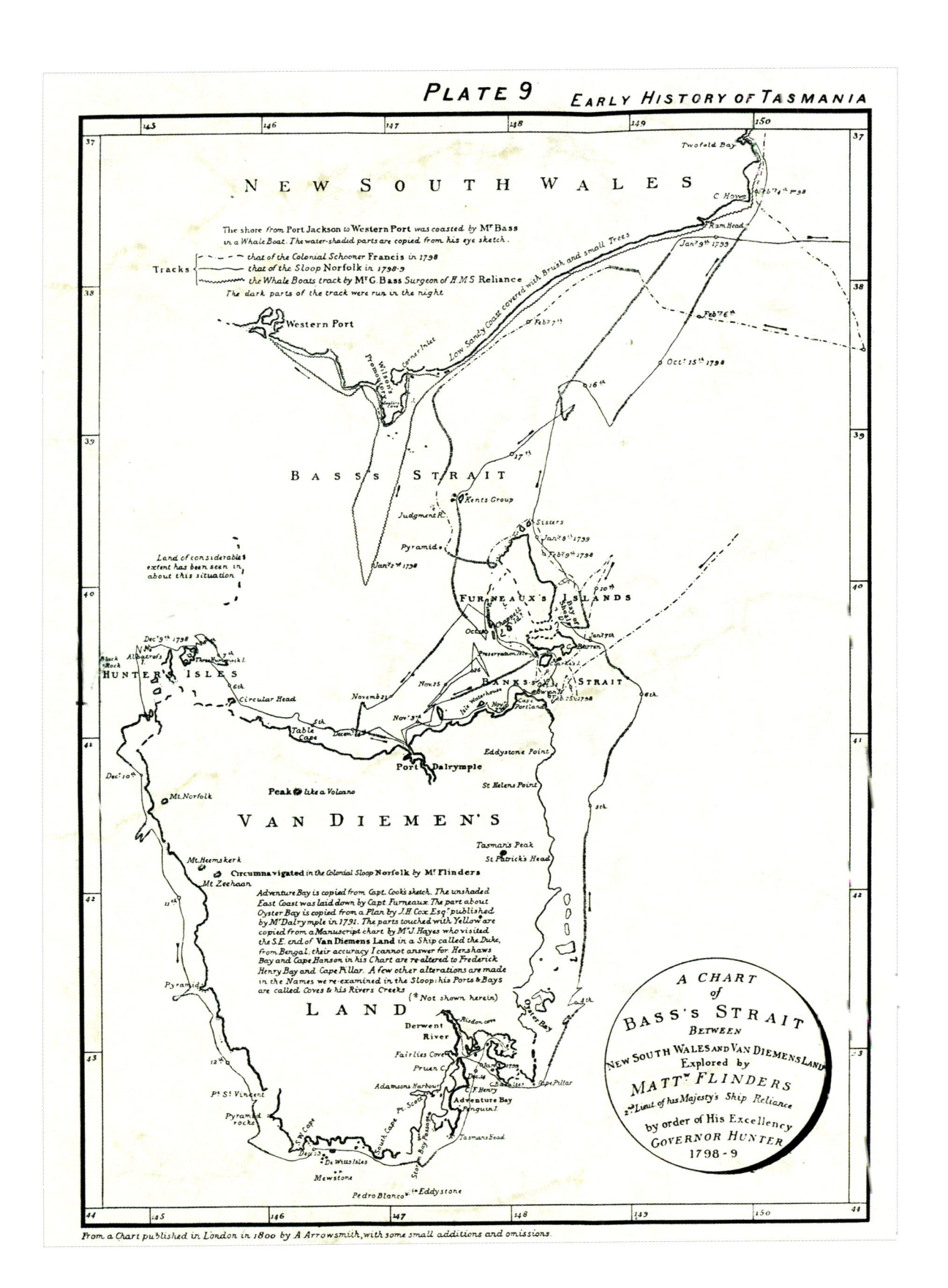

PLATE 9   EARLY HISTORY OF TASMANIA

The indigenous people of Van Diemen's Land extended quite a pleasant welcome to the explorers that made the first contacts with them, at the end of the 18th and beginning of the 19th centuries.
It is a great shame that their oral tradition could not be recorded as they would have been able to provide precious information on the numbers, the distribution and the habits of the thylacine, which they had undoubtedly lived alongside for centuries.

however the same man had already told us such obviously absurd tales about the savages living on the island and some sort of tiger found there, that his lies led us to believe everything else he had told us was pure invention".

John Dunmore told this story in his excellent book "*Les Explorateurs Français dans le Pacifique*"[1] (French Explorers in the Pacific), adding: "*And, in reality, unless this man had been frightened by another group of sealers or awakened by the raucous calls of sea elephants, he must simply be considered a liar.*"

This conclusion seems a little hasty. Indeed, we believe that the allusion to savages and some sort of tiger is not simply the fallacy of an unbridled imagination, but that the stowaway, totally unaware of where he was to make an accurate judgment of distances, had simply confused Campbell Island with Tasmania. Both islands are in fact more or less on the same course that the sailing ships of the day used to follow from Port Jackson (Sydney) to Cape Horn.

That the man himself had spent time in Tasmania is somewhat unlikely, for at the end of 1819, there were only about 500 convicts in Van Diemen's Land, as it was still known. To a certain extent these convicts were forerunners, given that the penitentiaries

CHAPTER FIVE — THE FIRST EYE-WITNESSES

at Macquarie Harbour and Port Arthur were only founded in 1822 and 1830 respectively. Perhaps the man had been one of the initial deportees, but it is unlikely that he himself would have hunted seals, as this was mainly done by free men[2]. In fact, evidence suggests that the man was making up stories, and in pretending to be an eye witness, he was only relating rumours going around Port Jackson about "some sort of tiger" encountered in Tasmania.

It is also possible that, with complete disregard for climatic conditions, he considered anything that was true for Tasmania was also true for Campbell Island, the name of which was already known to the inhabitants of Port Jackson. Discovered on 4 January 1810 by Captain Frederick Hasselburgh, Campbell Island was in fact immediately invaded by a multitude of sealing vessels, whose owners were generally prosperous and respected in the capital of New South Wales. The same was true for Macquarie Island, which was discovered in the same year.

Whatever the case may be, this story from one of the stowaways on *Uranie* was well worth relating, in so far as we consider it to be proof that even in 1819, the reputation of the "tiger" had already travelled beyond the confines of Tasmania, and was of concern to the inhabitants of the mainland.

Surveyor-General of the Colony at the time, Mr Evans owned properties in the northern parts of Tasmania, where he lived from 1809 to 1824. Evans travelled extensively and in 1822 commented on the "*opossum hyena*", stating that few of these had been seen. He was somewhat mistaken when he described the Tasmanian tiger as an animal of the panther tribe, adding that the species ravaged flocks infrequently yet extensively. A drawing of

A highly imaginative drawing of a Tasmanian tiger.

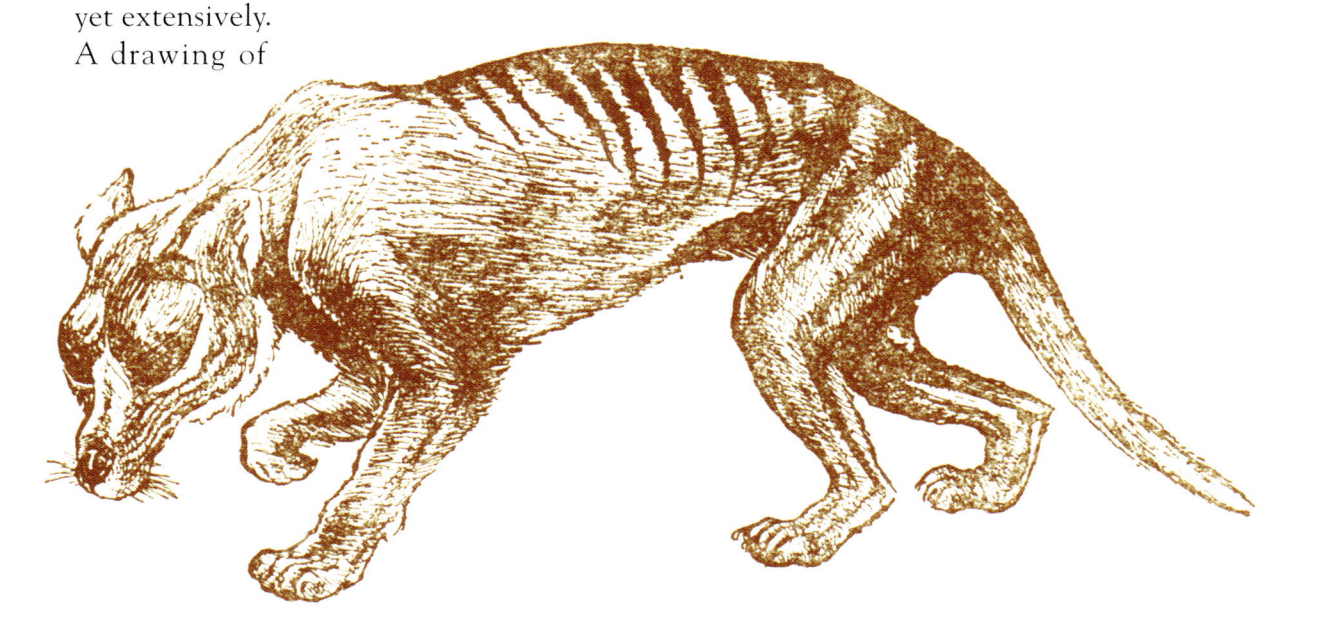

unknown origin shows an artist's impression of a most ferocious creature, and perhaps Evans' comments were partly responsible.

The scarcity of thylacines was the subject of comment in 1832, when it was stated that *"its habits are little known; indeed it is seldom, if ever seen, save when caught in the night in the snares set for it by the shepherd"*. This indicates that the tigers were perhaps shy of man because of centuries of persecution by Aborigines and trappers who set snares to specifically capture them as early as 1830. Being nocturnal, it was easy for a thylacine to stay out of sight, and in the early days country people did not venture out into the night except in dire circumstances, for fear of attack from bushrangers or Aborigines.

More information on the geographic distribution of the tigers comes from Widowson's 1829 account of the species *"frequenting the wilds of Van Diemen's Land and scarcely heard of in located districts"*. In 1829, Mudie stated the Tasmanian tiger existed in *"inland Tasmania"* but *"did not approach the thickly populated parts of the country"*. Both Widowson's and Mudie's reports concur in saying that the Tasmanian tiger lived in the more remote and largely unexplored parts of the colony that were not used for grazing or farming, however both express the same contradiction in saying that the tigers killed sheep on farms. The presence of farms certainly implies some degree of human settlement where the tigers made their kills, and could hardly be considered as "the wilds".

The first prediction of extinction came nearly forty years after the arrival of Europeans in Tasmania, when the thylacine was described as a *"species whose term of existence seems to be fast waning to its close"*. This 1842 report does not however specify

A primitive yet undoubtedly effective tiger trap built in 1823. The same basic design was used for traps in the 1960's.

whether the prediction is based on declining numbers or if it was made because the infrequent sightings of the species gave the impression that thylacine was vanishing. In a letter dated 12 December 1850, Gunn reported to the Zoological Society of London that the Tasmanian tiger was confined to the western mountain tops over 3,500 feet high, although he soon changed his mind. In Tasmania in the 1840-1860 period he recorded tigers *"occurring all over the island from mountain tops to sea level"*. This view of a widespread geographic distribution does not concur with the opinions of earlier writers or even Gunn's contemporaries. At the same time, there were other predictions of extinction for the species, which had become extremely rare, and 10 years later it was noted that the tigers were *"not very numerous and but seldom seen"*.

The statement that Tasmanian tigers inhabited only the mountain tops implies that the species had been driven from the more settled areas. By this time there seems to be some agreement that thylacine was confined to the remote parts and was not usually found in the settled parts of the country. Hence, extinction loomed for the species, given that sheep had become part of their diet soon after European settlement. It was obvious that any tiger unfortunate enough to be found on a sheep farm was unlikely to be left alone, and would be hunted down and killed.

It is clear, even at this early time, that a few far-sighted individuals were sounding the warning bells. By 1860 or thereabouts, it was becoming clear that the Tasmanian tiger was recognised as scarce, even rare, and in this day and age would be classed as a threatened if not endangered species. Some tigers had developed a new habit which brought the whole species into disrepute - they killed sheep.

---

1 - Volume 2 - Chapter XIX - Sté Nouvelle des Editions du Pacifique - Tahiti - 1980.
2 - This is a general rule of thumb, but there were exceptions. In his book *"First Visitors"* which dates back to 1797-1810, J.S. Cumpston points out that escaped convicts did hunt for seal, particularly in the Clarke Island and Cape Barren region.

TASMANIAN TIGER — A LESSON TO BE LEARNT

Chapter Six
Thylacine an

d the artists

The abundance of illustrations relating to the thylacine is confirmation of the very real fascination that naturalists, from the very beginning, have had for the animal.

From the year 1808 when George Harris drew the first picture, a rather mediocre one at that, through to the popularisation of photography in the last two decades of the 19th century, the representations of this strange animal, some drawn, some painted, are countless.

Some of these, as the illustrations that follow prove, are pure fantasy. Without doubt one of the main reasons for this is the fact that the artists most often based their work on piecemeal information. Furthermore, they had very little to fear from critics, given that, at the time, Tasmania was very much a world away for Europeans, and information, be it scientific or otherwise, took much time to circulate.

The vision that these artists - often the naturalists themselves - had of the thylacine was all too frequently influenced by the images of animals that were already familiar to them, such as the dog, the cat, and to a lesser extent, the wolf. The only touch of originality comes from the stripes. And since the "tiger" of the Antipodes had already forged for itself a reputation as a vicious exterminator of flocks of sheep, every one of these artists undertook to give the thylacine an air of cruelty that, in the end, is not particularly representative - as we shall see later in observing specimens held in captivity - of the animal's true personality.

It wasn't until the year 1863 that John Gould, an attentive observer and an artist of significant talent, provided a first and almost perfect image of the largest predator ever

TASMANIAN TIGER — A LESSON TO BE LEARNT

Previous page:
This colour engraving figures in the work entitled *Animal Kingdom - The Class Mammalia* by the famous French naturalist, baron Cuvier (London, Whittaker, 1827). Our thylacine, whose stripes are illustrated in an exceptionally discreet fashion, is shown with fur and features that are not unlike those of a bear cub!
Below:
It is difficult to give an exact date for this drawing from the Frenchman Pourrat. We do know that the name *Thylacinus Harrisii*, given to the animal in 1824, disappeared in 1831. The position of the hind legs, shown here flexed, are similar to that of a basset hound. The head is also a total failure with the jaw too weak, cat's eyes and the ears out of proportion.

to have lived in Australia and its offshore island Tasmania.

This did not however close the door on certain forms of graphic delirium. Thylacines chasing… kiwis (which have only ever existed in New Zealand) or feeding on a platypus (how on earth could the thylacine have captured it?) of outlandish proportions!

In the final analysis, all these fantasies do make for a charming kaleidoscope and are in no way exceptional. One need only see to what extent explorers of the 17th century gave their contemporaries deformed visions of landscapes, animals and even fruit they discovered as they travelled throughout the New World, to understand that a certain desire to strike imaginations, or make their stories all the more epic, when it wasn't simply a lack of ability to faithfully represent reality, incited them to take liberties with the latter.

No doubt the reader will take real pleasure in observing the illustrated pages that follow, such is the exotic aura surrounding them. The illustrations bear witness to the amazement that our recent ancestors felt as they lifted the veil concealing the last mysteries of our planet.

CHAPTER SIX — THYLACINE AND THE ARTISTS

1820's

Above:
The first representation of the Tasmanian tiger in a German treatise of natural history. We owe it to the hand of a certain Schinz, and it dates from 1827. The tail appears too long and is also badly joined to the animal's hindquarters. As for the look, it resembles more a rodent with its round, candid eyes and ears that are too rounded. It is obvious that the artist had no model at his disposal and based his work on descriptions he had read.

Next spread:
This fine engraving is full of imagination, but clearly lacks accuracy. The body and the head are too tapered, the tail is too short and the dentition is quite simply frightening (!), without doubt on purpose. This illustration was presented to the Linnean Society of London in March 1821 by Mr W. Lister Parker. It is a work by John William Lewin, a painter and drawer of animals born in England in 1770. Arriving in Australia aboard the *HMS Buffalo* sometime in 1800, he was a forerunner in his genre and his *Birds of NSW* met with great success. He was also the author of a grand painting entitled *Fish of Australia*. This representation of the Tasmanian tiger had never been published. It is not known what model the artist used. We do know that he never set foot in Tasmania. His only attempt to get there was a failure the *Lady Nelson*, the small vessel aboard which he was travelling to the island, was caught in a storm as it began the Bass Strait crossing, and had to turn back!

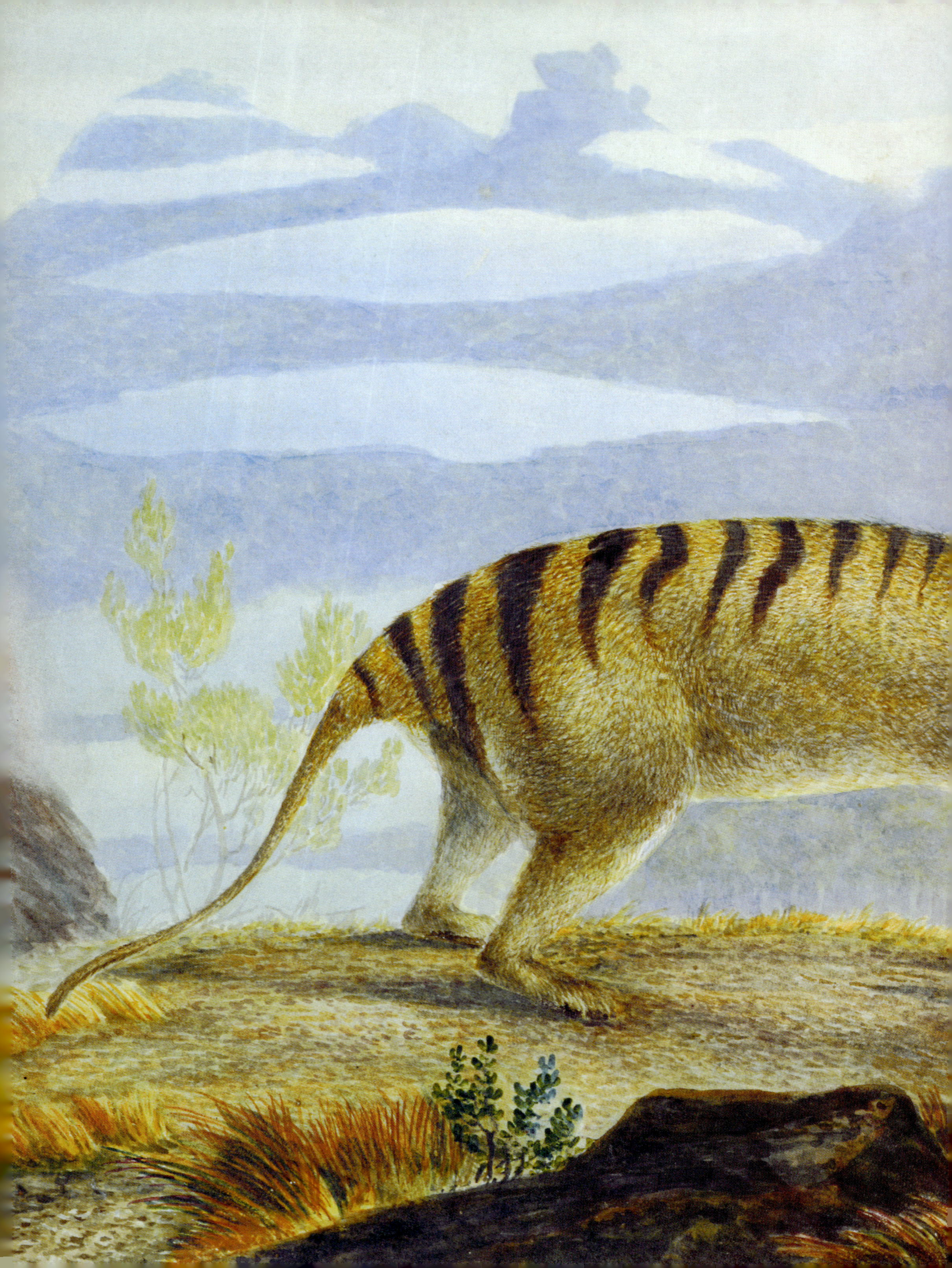

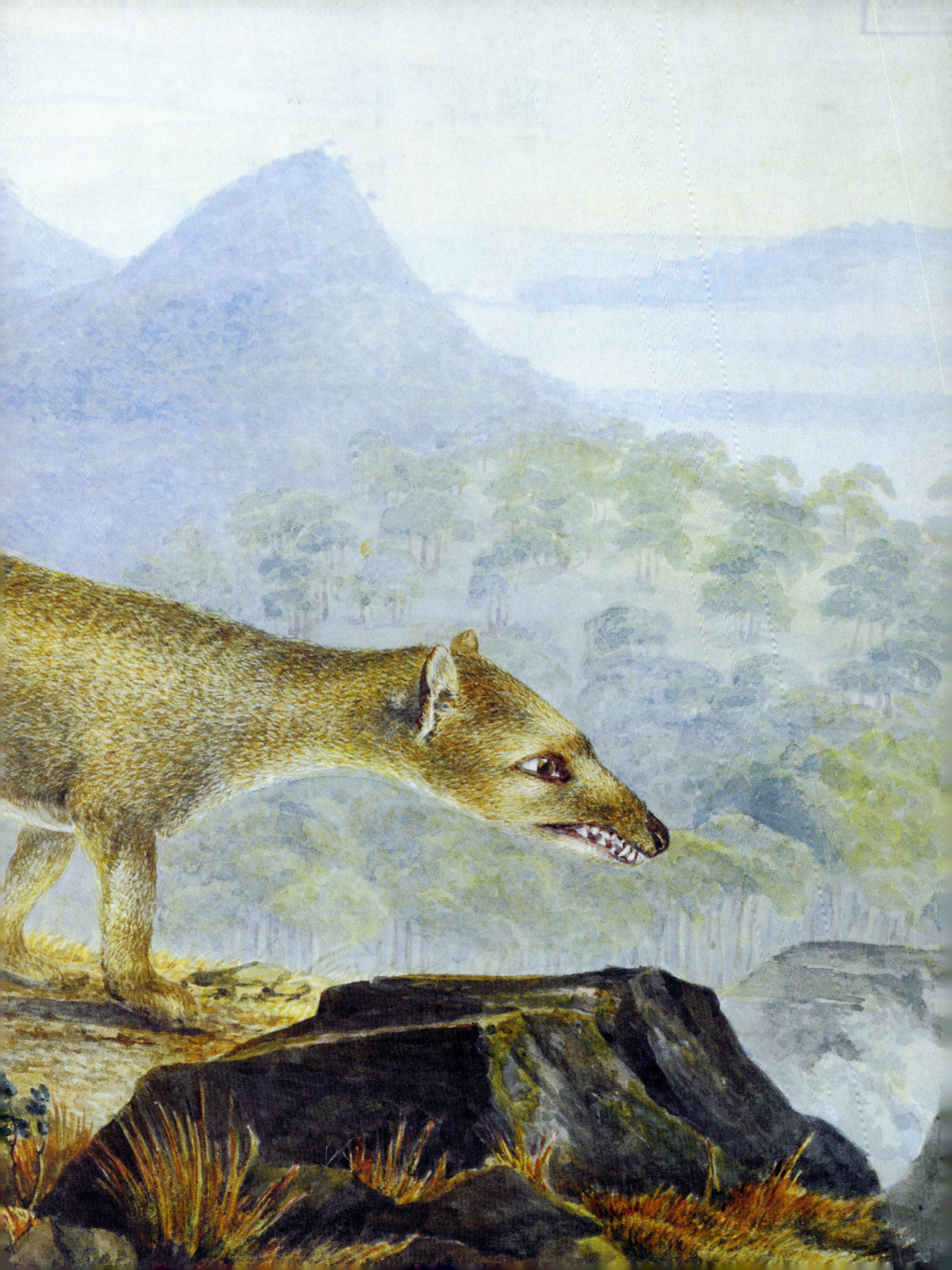

TASMANIAN TIGER — A LESSON TO BE LEARNT

Below: This engraving is not dated and the author remains unknown. It is only of passing interest due to the fact that, although the body proportions are more or less exact, the tail is too tapered. The incorrect curve of the back could also be criticised, along with the length of the head, the ridiculously pointed nose, the muzzle and bottom jaw incorrectly represented and, above all, the eyes, which are too small and devoid of expression.

Opposite above: Another French representation, this one dating from the year 1839. The tail is out of proportion and the body too slender. The angelic air hardly reflects the ferocious character of animal. On the other hand, the proportions of the legs are close to reality.

Opposite below: A Waterhouse engraving, published for the first time in 1841 and again in 1846. It is a drawing of total fantasy, which is hardly excusable given that, by this time, the silhouette of the Tasmanian tiger was becoming well known to naturalists in both the new and old worlds. The animal is shown hunch-backed, with the appearance of a big rat bearing a plume at the end of its tail! As for the head, it looks very much like that of a nasty little lap-dog.

1835-1846

CHAPTER SIX — THYLACINE AND THE ARTISTS

TASMANIAN TIGER — A LESSON TO BE LEARNT

1850's

Below:
This illustration of a pair of tigers dates back to 1850. It appeared in the newsletter of the *Royal Zoological Society of London*. The male is in a position of observation, and resembles the image given by E. Perceval Wright (who probably found inspiration in this illustration) in 1892. The female, shown face on and lying down, has an angelic look about her. Her ears, a little too rounded, also lack straightness.

Left:
In the years 1857 - 1858, the *London Zoo Guide* took great pride in presenting, among other rarities, a Tasmanian tiger. The animal figures in the catalogue given to visitors of this period. Only a silhouette is shown, but it is very accurate.

Right:
In their desire to show several species from the Antipodes in a single drawing, the artists of the mid-19th century frequently took liberties with reality. Such is the case of this German engraving where the pair of tigers, whose forward parts of the body are too big, are hunting a group of kangaroos (which is plausible) as well as an emu (which is a lot less likely). Indeed, this species disappeared very early from the Tasmanian scene, decimated by the first settlers from the beginning of the 19th century. Furthermore, it would have been no mean feat for animals the size of our "tiger" to catch and put to ground an emu, before feasting on it. On the other hand, the representation of the vegetation is interesting.

CHAPTER SIX — THYLACINE AND THE ARTISTS

TASMANIAN TIGER — A LESSON TO BE LEARNT

1863

CHAPTER SIX — THYLACINE AND THE ARTISTS

These famous drawings, which illustrate the masterly works of Gould, date back to 1863. They are nearly perfect, both in terms of accuracy and artistry. If one were absolutely determined to make criticisms, however, it could be pointed out that, although the geometry of their bodies is well represented, the animals do seem a little too corpulent. As for the close up of the head, shown separately, it reveals a muzzle that is slightly too tapered. The eyes, almost too expressive, also show a somewhat mysterious cruelty that we do not find in the photos that would be taken, a few years later, of the animal in captivity.

TASMANIAN TIGER — A LESSON TO BE LEARNT

Below: This 1864 etching by Harriet Scott became widely published. There is no dispute that it constitutes the best representation of the tail of the thylacine. The body proportions are also excellent, but the muzzle is a little too flattened; the head also lacks power. It is worth noting the attention to detail shown in the representation of the plant life and landscape in the background.

Opposite above: Drawing by the German Kretschner published for the first time in 1865. The resemblance can be described as good, despite the artist obviously being influenced by the wolf, in particular with regard to the position of the legs and the shape of the head and the ears.

Opposite below: Another drawing that comes to us from Germany. The artist's name, who drew it in 1877, was Mutzel. It is a very good representation of a pair of thylacines, one lying down, the other standing up. The proportions can be described as irreproachable.

CHAPTER SIX — THYLACINE AND THE ARTISTS

1865 to 1877

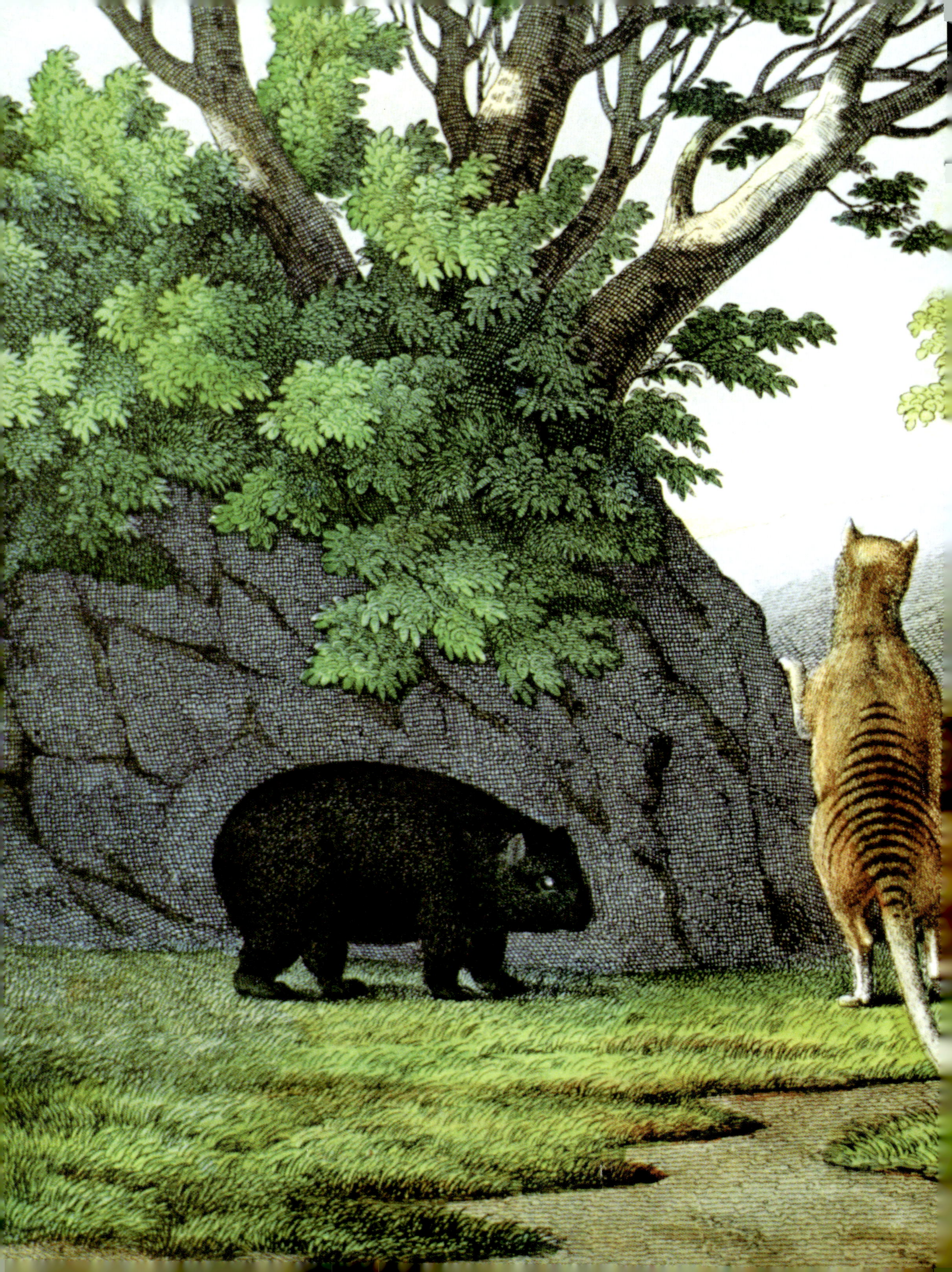

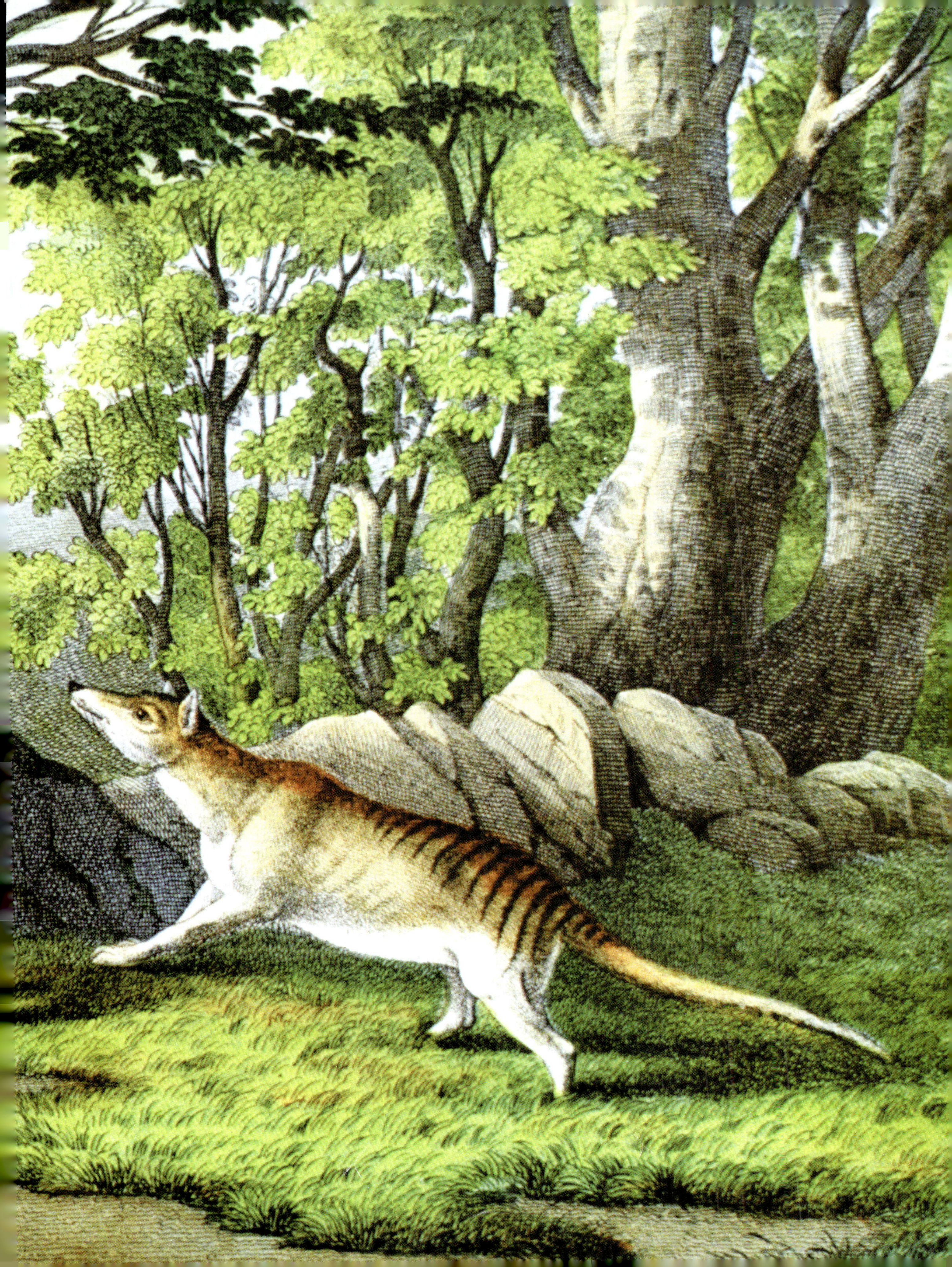

TASMANIAN TIGER — A LESSON TO BE LEARNT

1880's

Previous spread:
We are coming to the end of the 19th century and the *Beutelwolf* continues to fascinate the Germans. Here, the artist Schreiber shows a pair with a wombat (against all probability, as the latter is nocturnal, in the first instance, and was, without doubt, a choice prey for thylacines). The two predators are faithfully represented, although a little pot bellied. The artist went to great pains to show the female's pouch (that he perhaps wished to illustrate full) whilst the male, its front feet upon a rock, observes the nearby valley in the hope of sighting prey.

Above:
In the natural history treatise from which this drawing is taken, the two "tigers" shown take on the appearance of large tomcats with excessively thick tails. The head is very poorly represented. Again, it is the stripes on the fur, on which the attention of all artists seems to have focused over the years, that are the most accurate.

CHAPTER SIX — THYLACINE AND THE ARTISTS

Above left: The German E. Specht, author of this drawing, which in itself is quite good, obviously did not know that the apteryx (more commonly known as the kiwi), has only ever existed in New Zealand!

Above right: Still in Germany with this drawing by Kuhnert, dated 1890. The kangaroo under chase, without doubt a Forester judging by its large size, is remarkably well represented. Its chasers, whose position is decidedly feline, are a little less well done.

Left: Another fine engraving showing a pair of tigers on the heels of an emu. All three animals are quite accurately represented. Only the size of the emu, compared to that of the animals pestering it, gives rise to reservations, as the Tasmanian emu was not as big as its counterparts living on the mainland.

1890's

TASMANIAN TIGER — A LESSON TO BE LEARNT

CHAPTER SIX — THYLACINE AND THE ARTISTS

Opposite above: This work by E. Perceval Wright (1892) does nothing to enrich the subject. The proportions of the head are bad and the thylacine has never shown its fur raised in such a manner on its back and neck. The commentary accompanying this representation of the animal is worth reporting: "Thylacinus cynocephalus is the only species of the genus. It is the strongest and fiercest of all marsupials. It was formerly common in Tasmania where it has often been compared to the wolf, as it is about the same size and has the same sanguinary appetite as the animal. Like the wolf it frequently falls upon flocks of sheep, which offer it easy prey. Very common along the coast it lives principally, it is said, on animals remains thrown up by the sea on the shore; it also eats crabs."

Opposite below: With this composition, we enter into the domain of fiction. It is indeed highly unlikely that the platypus, so lacking in flesh and covered with thick fur, would ever have been among the species preyed upon by the Tasmanian tiger. The tiger, on the other hand, is extremely well illustrated. It is a male, whose testicles are exactly the same as those that can be seen in the photograph of an animal living in captivity (page 51). Such accuracy is all the more surprising given that the scale is fanciful, the famished tiger taking on a platypus that is almost as big as itself!

Below: The Royal Natural History by Richard Lydekker (London, 1894) includes this illustration of a thylacine on the lookout. The document is not devoid of interest, even though the animal's neck is again shown too thick.

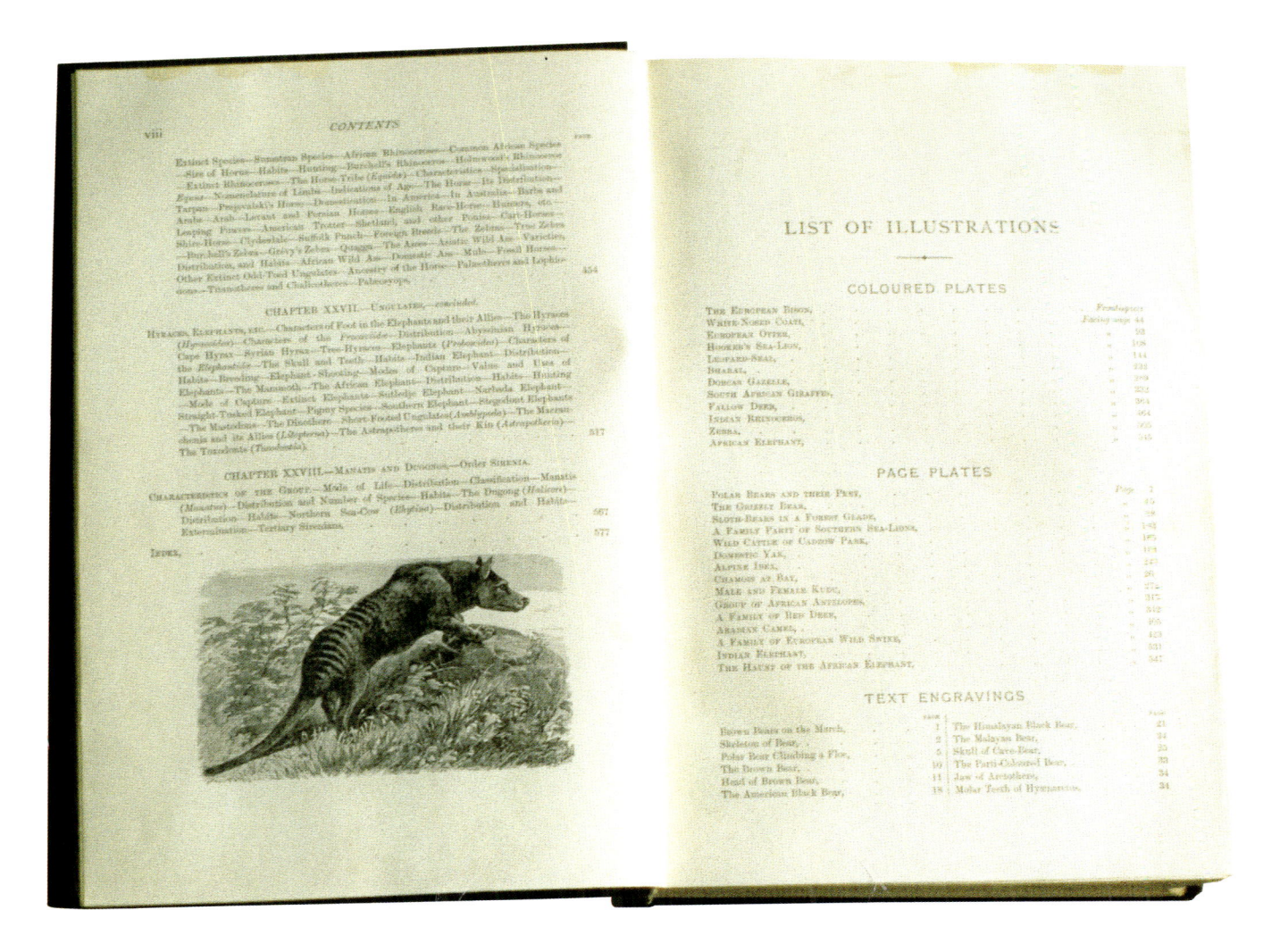

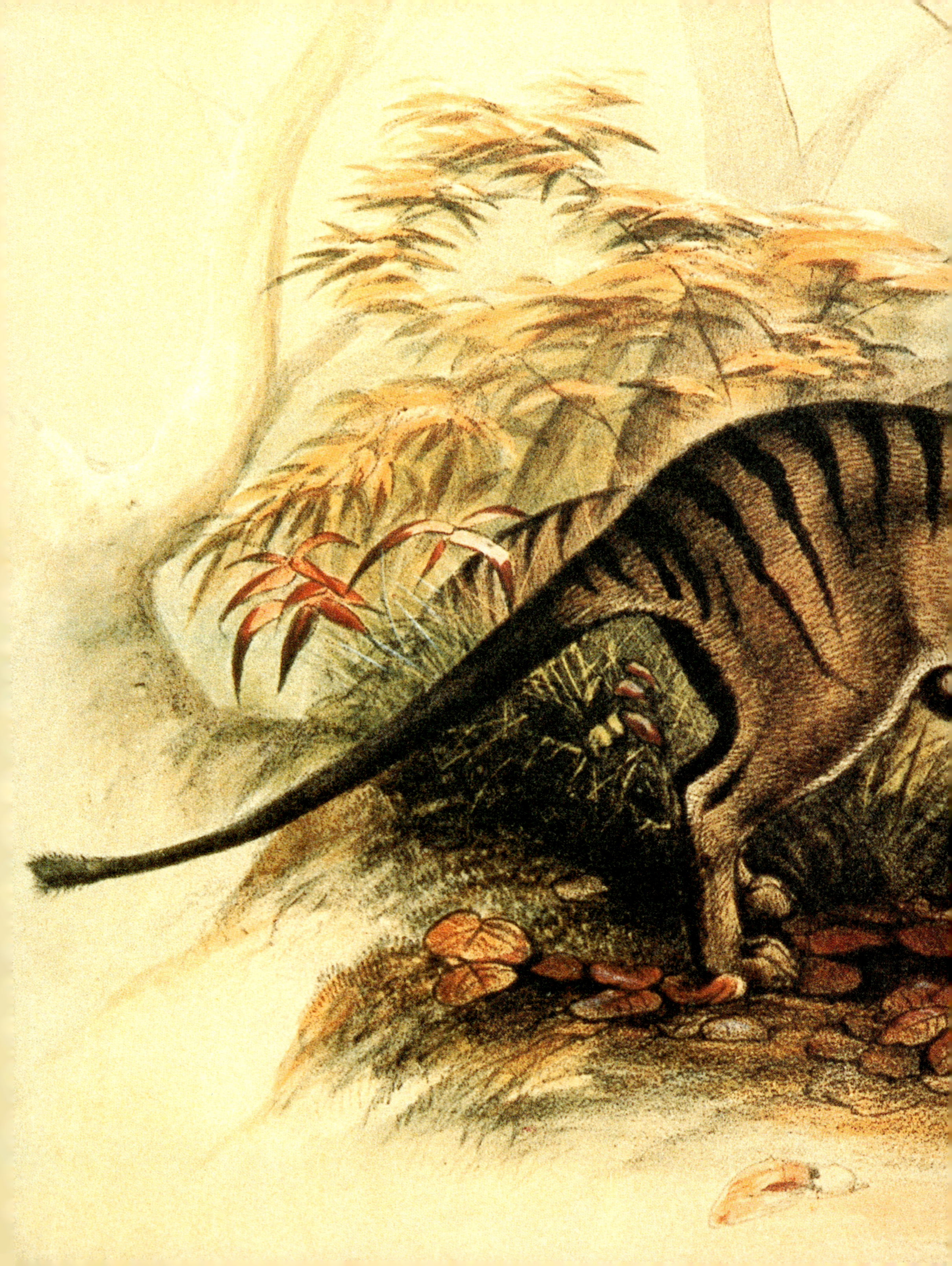

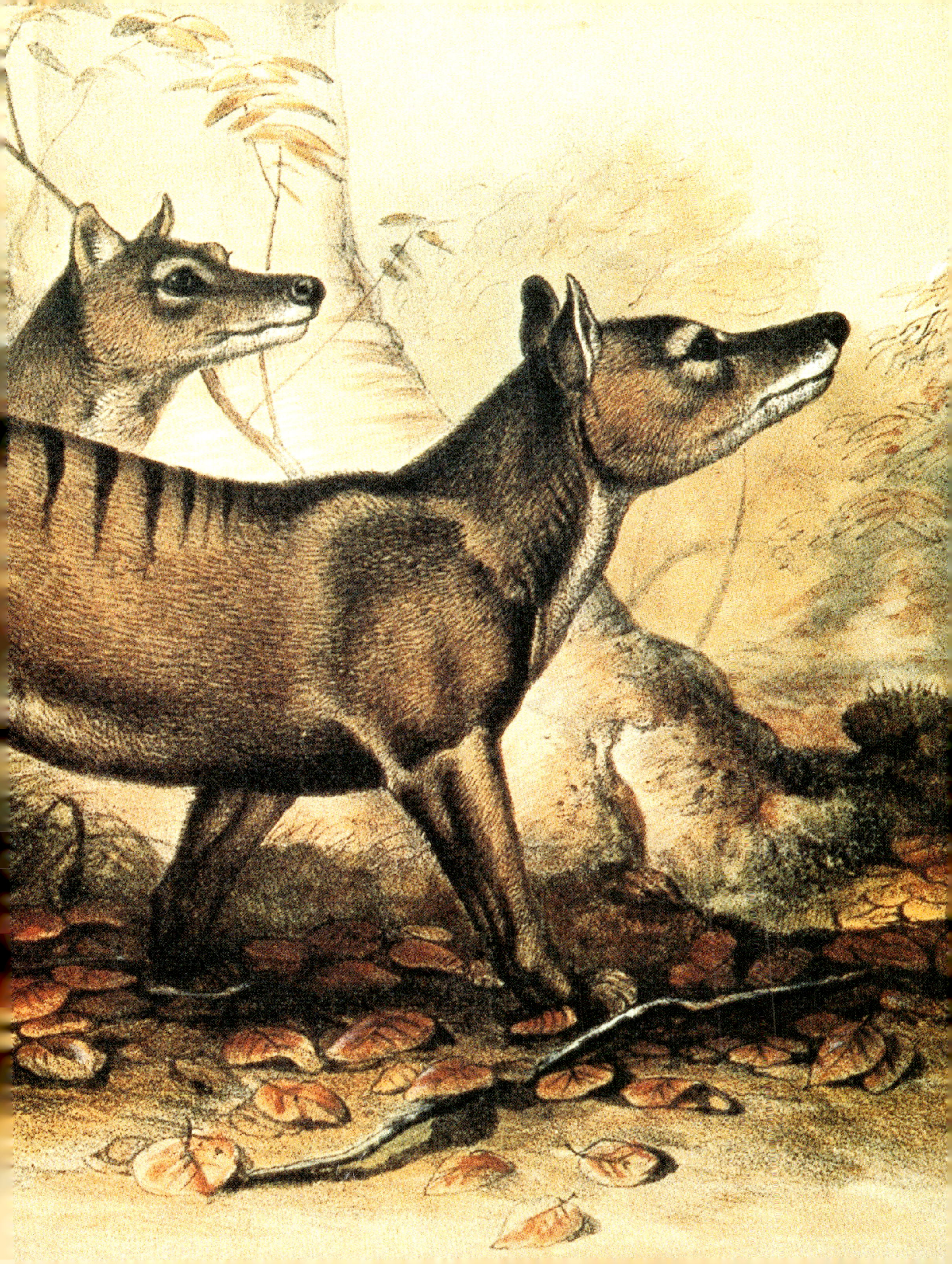

TASMANIAN TIGER — A LESSON TO BE LEARNT

1900's

Previous spead:
A fine representation of a pair of thylacines in their natural surroundings, anonymous and undated. The proportions are excellent. One could only just criticise the muzzle, a little too flattened, and the plume adorning the end of the tail.

Below:
This drawing by Lankaster (1902) is of a lesser quality than the previous ones. The limbs appear over-sized and the animal, with its squat body, gives an exaggerated impression of strength.

CHAPTER SIX — THYLACINE AND THE ARTISTS

The 20th century was fifteen years old when Lydekker, after some many others, went about publishing this representation of the thylacine. The result is somewhat disappointing. Although the body of the animal about to leap is correctly proportioned, the neck again seems over-sized.

TASMANIAN TIGER — A LESSON TO BE LEARNT

Chapter Seven
and bount

Persecution by hunting

There is no doubt thylacines soon learned that sheep were easy prey, and quickly adapted to exploiting this new source of food. Sheep had arrived with the first group of settlers in 1803, to form the basis of the vital food and clothing industries, and there are many references to thylacine attacks on sheep. There were complaints of a panther-like animal causing dreadful havoc among the flocks and the number of lambs killed by Tasmanian tigers, and it was said they were even raiding poultry yards. At George's Bay on the East Coast there were also reports of tigers killing sheep. An entry in George Robinson's diary on 17 August 1830 reads *"great numbers of hyenas killed sheep at Surrey Hills"*.

The Van Diemen's Land Company records are the most reliable source of information regarding the Tasmanian tiger's early relationship with sheep. This Royal Charter Company was established in 1824 to develop agricultural and other interests in Tasmania. It was a big operation and meticulous records were kept on its activities. Each Station kept a diary in which all the daily work was entered, along with the monthly stock returns which provide valuable information on the Tasmanian tiger and its effect on the flocks. The company had vast holdings, including 100,000 acres at Woolnorth in the north-west end of the State, with smaller properties in other locations such as Hampshire.

Hyenas, dogs, vagabonds and Aborigines were all sources of problems. In 1830 the Van Diemen's Land Company introduced a bounty scheme, in the hope of ridding its properties of dogs, devils and Tasmanian tigers. The offer read: "*five shillings for every male hyena, seven shillings for every female hyena (with or without young). Half the above prices for male and female devils and wild dogs. When 20 hyenas have been destroyed the reward for the next 20 will be increased to six shillings and eight shillings respectively and afterwards an additional shilling per head will be made after every seven killed until the reward makes 10 shillings for every male and 12 shillings for every female. A proportionate amount will in like manner be made to the rewards given for devils and wild dogs*".

The major thrust of the bounty scheme was aimed at the Tasmanian tiger. The increasing scale of the rewards shows that the Company was quite determined to be rid of these pests, obviously present in considerable numbers on its properties, and realised that a major effort would be required to reduce the population. The dogs and devils were not regarded as such a serious threat, although more numerous, they were easier to destroy and were only worth half a thylacine as far as the bounty rewards were concerned.

Bounty payment records have unfortunately disappeared, and only two transactions have come to light. The first is a seven shilling payment to an assigned servant, McKay, for a bitch hyena killed at Epping Forest near Hampshire on 22 August 1830. The second bounty payment was for a dead hyena found in 1832 at Mt Cameron. The animal had been killed by natives, skinned and the pelt taken to Cape Grim to "*get 10 shillings from the Company*". The amount of this particular reward implies that the Company was paying the maximum amount, so it can be concluded that at least forty-seven tigers had been killed since the inauguration of the scheme in 1830. The Van Diemen's Land Company in fact had more trouble with wild dogs than it did with thylacines. In 1834, Backhouse reported that wild dogs were very numerous in the Emu Bay area, and that they killed many sheep.

Previous page: **A successful hunter at Maydena, 1912.**
Right: **Government Gazette notification of the Bounty scheme.**

The bounty scheme would have generated much fervour and must have resulted in financial gain for employees on properties where there were large numbers of devils and thylacines. What is more, the rewards were very generous for the time. In terms of bounty payments, sexual discrimination was removed in 1840 with the basic reward increasing to 6 shillings per scalp, 8 shillings for between ten and twenty scalps, and ten shillings for more than twenty. The Company must have been having great difficulty with the animals for such bounty to be offered. To further combat the high sheep losses on the Woolnorth property, a company-appointed trapper became the "tiger man", a job that existed until about 1910. The "tiger man" was given a hut and his keep, and spent most of his time trapping, acting as a shepherd or assistant as the need arose. He lived at Mt Cameron West, visiting the homestead from time to time with skins and dead tigers which he exchanged for supplies. Such visits always rated an entry in the Station diary.

The bounty scheme and the appointing of a trapper certainly indicate that the Company was having trouble with thylacines killing sheep during the early stages of the property's development. There are no comments or statistics about this in the few remaining records, and in fact the only statistics are quite to the contrary. Stock records for Woolnorth between 1832 and 1834 show no losses due to Tasmanian tigers, nor is there any comment on the topic. Yet the records are full of all sorts of calamities, including between 80 and 150 accidents every year. Sheep were being killed by tigers in 1839, however the numbers are not known. By 1843 the losses had increased greatly, and something needed to be done about it.

Snaring was the typical method used to catch a thylacine. A "necker" snare was most common, consisting of a noose of wire or hemp placed around holes in fences or constricted passages along animal trails. The necker was placed in such a manner as to strangle an animal when it was caught in the noose. A paling was often removed from fences, allowing the predator to pass through, at the same time providing an ideal snaring site. Portions of this old fencing were still around in the early sixties, particularly on the Three Sticks run in the shelter of tea-trees. It is doubtful that treadle or springer type

Reward for Destruction of Native Tigers.

154. A reward of One Pound shall be payable out of the Consolidated Revenue for the destruction of every full-grown Native Tiger (*Thylacinus cynocephalus*), and the sum of Ten Shillings for every half-grown or young Native Tiger, subject to the following conditions.

155. The person claiming such reward, or person authorised in writing by him, shall produce to the proper officer the skin of the animal complete, with head (or scalp) and paws adhering thereto, and shall satisfy the said officer that the said animal was captured and destroyed.

156. Any one of the following officers is hereby authorised to certify to the destruction of Native Tigers, and upon his certificate the amount of reward shall be paid:—

> The Secretary for Lands,
> A Stipendiary Magistrate,
> A Warden of a Municipality,
> A Police Clerk,
> A Council Clerk,
> A Superintendent of Police, and
> A sub-Inspector of Police.

157. Upon the production of a skin so complete as aforesaid to any such officer, he shall make a round hole therein immediately behind the forearm, not less than half an inch in diameter, and the skin so mutilated shall become the property of the person claiming the reward.

158. The officer to whom the claim for reward is made shall declare upon the voucher for payment as follows:—

> I hereby certify that .... has satisfied me that .... has destroyed * .... Native Tiger... the complete skin... of which ha...... been produced to me, and that I have mutilated the same, according to law.

159. On and after the day on which these Regulations shall come into force, no reward shall be paid for the production of a tiger skin unless the same shall be produced in a complete condition, as hereinbefore provided.

160. All claims for rewards for the destruction of Tigers, together with voucher for payment, duly signed and certified to, are to be forwarded to the Secretary for Lands, Hobart.

snares were used, mainly because there were few good springers to be had for miles around. Tigers caught in necker snares would have often been dead or very battered when found, although a few were healthy enough to be sent to zoos in later times. Wainwright was the last "tiger man", and he sent three specimens to Hobart Zoo in 1913 or 1914. Based at Mt Cameron West, Wainwright operated all over the property as the need arose.

The economic pressures of the period led the Company's manager to conclude that the removal of thylacines would result in a substantial drop in the numbers of sheep killed. He did not realise that as soon as a tiger was killed another moved in to take its place. The inaccuracy of this conclusion is apparent from Table 7.1[1], in which dogs feature as the major killers, however it still meant that Tasmanian tigers were continuously slaughtered on the property, although many of them would probably never have killed a sheep at all.

The Van Diemen's Land Company also experienced sheep losses on other properties. The Surrey Hills Station was situated about 45 kilometres inland from Burnie on the north-west coast. As early as 1829 the choice and value of the land was questioned. Bullocks died in winter and the land was found unsuitable for stock even in the summer months. The number of predator kills can be calculated from the Surrey Hills Station diaries. In the period from 1832 to 1849, a total of 903 sheep were killed by predators; 146 kills were blamed on thylacines, 299 attributed to dogs and the remaining 458 killed by either dogs or thylacines. Apart from the death of a single lamb, no kills

CHAPTER SEVEN — PERSECUTION AND BOUNTY HUNTING

The introduction of sheep in Tasmania by the first European settlers was a godsend for the thylacines, who had no hesitation in taking their "cut" among the flocks of these vulnerable animals with such tasty flesh. This German engraving dates back to the first half of the 19th century, and shows us our predator in action. A lamb has just been taken down by one of the pair of pursuing tigers, while the rest of the flock escapes before the panic-stricken eyes of the shepherd.

were attributed to the Tasmanian devil. The number of sheep killed increased around 1833-34, and it is clear that dogs were the more serious predator.

At both Woolnorth and Surrey Hills, the predatory canines were of the large cattle-dog type, that had either gone wild or were part of established bands of Aborigines or vagabonds, the latter probably being escaped convicts living by their wits. Both Aborigines and vagabonds were blamed for the theft of sheep on the Company's properties.

In addition to losses to predators, sheep also died from disease, harsh weather, theft and accidents. The total losses from all causes on the Woolnorth property amounted to 5,941 for the period between 1840 and 1848. The average annual flock on the property amounted to about 6,800 sheep, and at times the losses were as high as 16-17%, which would have been hard to sustain for any great period of time. In this context, it is not at all surprising that the Company made every effort to reduce losses by trying to control the predators.

The percentage of sheep killed is considerably higher for Surrey Hills compared to Woolnorth. In 1833, a total 394 sheep were lost, and another 206 the following year. With a total flock of 1,800 sheep in January 1834, annual losses to predators can be estimated at between 10 and 20%, proportions that are certainly high enough to justify some sort of control measures. The Company responded by yarding the sheep every night, but in the end was obliged to remove sheep from the property. In July 1834, 500 sheep were taken off the Surrey Hills property and the mortality rate dropped dramatically. The removal policy continued until there were only 50 sheep left in October of that year. The casualty rate remained low and the few kills that did occur were attributed to dogs.

The number of sheep was increased by 200 the following February, and more were put on the land around 1839-40 when predation was kept to a minimum through nightly yarding of stock. Predator activity increased sharply in 1845-46, only to fall away again. No further losses were experienced through to 1852, at which time the records end. Perhaps the most exciting evening of all occurred in 1844 when a Tasmanian tiger killed a pig, with all the commotion this would have caused in the middle of a 19th century night.

There was more justification for the tiger man's employ at Surrey Hills and Hampshire than on Woolnorth, however there is no evidence in the diaries nor from other sources that indicates any other action, save the not very positive policy of removing the sheep from the land. There is no trace of the tiger man ever being loaned from Woolnorth, and there is no mention of snaring activity and capture of thylacines at Surrey Hills and Hampshire. There are no other early records of Tasmanian tigers killing sheep, nor of the effects on flocks. With the exception of the Van Diemen's Land Company in 1874, all subsequent estimates of thylacine and sheep numbers are largely based on hearsay, some of which is highly suspect.

Right: **Distribution of payments for 2,131 thylacines during the bounty period, 1888-1909.**

It is impossible to estimate how many thylacines were involved in the killing of sheep, and even more so to evaluate the numbers of animals. Several trappers, among them T. Pearce, have said that only a few tigers became sheep killers, ranging down from the forests to attack the sheep in the adjoining pastures. Some tigers became addictive killers, and Pearce tells of a night he saw one thylacine kill sixteen sheep. Other trappers have spoken of three and four kills in a single night. This sort of habit is more typical of the destructive killing of packs of dogs. Pearce was nonetheless emphatic in stating that many thylacines ignored sheep and would pass through a flock without paying any attention to them.

It is equally impossible to estimate the numbers of sheep lost to thylacine attacks. Few records are available and farmers would make extravagant claims, but there is absolutely no doubt that sheep losses were sufficiently high to force farmers into protecting their flocks against thylacine attacks. There are accounts, such as Bunce's in 1857, of guards being placed around the midlands flocks at night during the lambing season, and tales of thylacine being chased away from flocks on Woolnorth station.

In the 1870's, the comment was made that tigers that had previously attacked flocks were now confined to the mountains. The accuracy of these statements is open to

CHAPTER SEVEN — PERSECUTION AND BOUNTY HUNTING

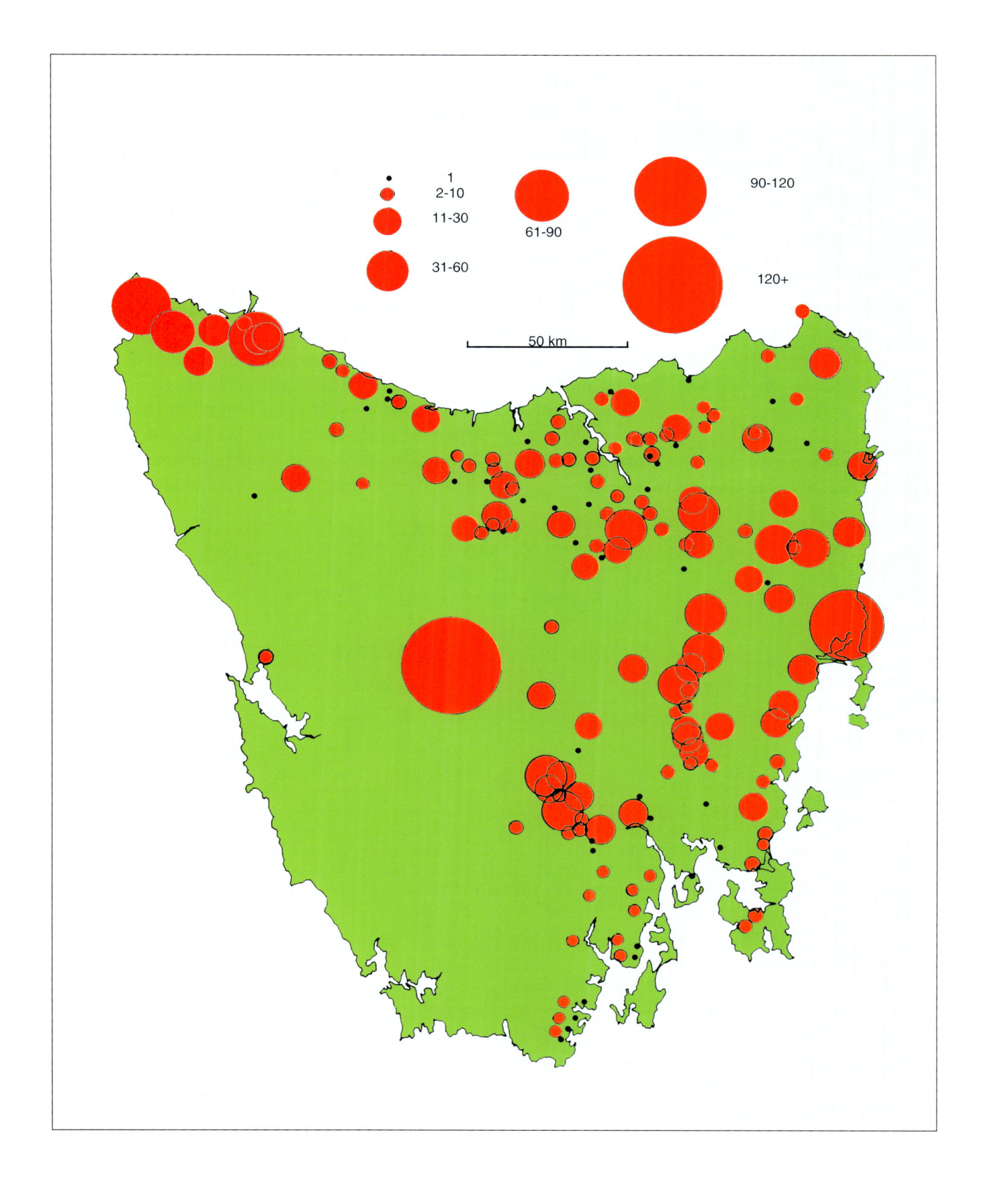

question, as there is plenty of evidence from the central highlands and the east coast that the attacks continued unabated. A property at Blessington allegedly lost 3,700 sheep to Tasmanian tigers between 1865 and 1870, and 448 sheep were killed in the same manner on this property in 1880. In 1953, F. Burbury wrote that his property lost 700 sheep from a flock of 2,000 at Lake Tooms in a single year around this time. A 1927 account stated that losses on the Burbury property at "the Island" amounted to 450, 350 and 300 in consecutive years.

Other property owners, such as Mr Morrison of St Peter's Pass, expected *"a hundred or so to be killed"*, with much higher losses in the summer pastures of the central highlands where the corpses were eaten by devils, making it difficult to estimate the number of sheep killed. The theft of sheep has always been a part of rural affairs in Tasmania, and many of the owners of the large hill runs may not have been aware of the extent of their losses due to theft. Several members of the Pearce family stated that losses through theft were much greater than those suffered through thylacine attacks. Not all of the hill runs suffered heavy losses, and W. Padman saw nine Tasmanian tigers over a single summer, but only three sheep were killed. Even today, the major

Market price of Tasmanian tiger skins, 1889. Although noted as scarce, the price seems to be very low.

FAUNA AND ANIMAL PRODUCTS.

STATISTICS RELATING TO ANIMAL PRODUCTS.

Fur Trade.

Ruling prices at the date of the following:—

		Undressed.	Dressed.
Wallaby Skins	per doz.	3s. to 6s.	9s. to 12s.
Kangaroo Skins	per doz.	12s. to 24s.	20s. to 32s.
Opossum (Black)	per doz.	18s. to 60s.	24s. to 66s.
Ditto (Grey)	per doz.	4s. to 12s.	10s. to 18s.
Tiger Cat Skins	per doz.	6s. to 24s.	12s. to 36s.
Native Cat Skins (Black)	per doz.	3s. to 9s.	7s. to 14s.
Ditto (Grey)	per doz.	1s. to 4s.	5s. to 10s.
Native Tiger Skins (scarce)	per doz.	36s. to 72s.	48s. to 84s.
*Rabbit Skins	per 1000	20s. to 27s. 6d.	
†Opossum Rugs, unfinished (Black)		£6 to £12 each	
Ditto (Grey)		£3 to £5 each	
Ditto, finished (Black)		£9 to £30 each	
Ditto (Grey)		£4 10s. to £12 each	

* Rabbit Skins range from Kittens, valued at about ¼d. each, to first-class heavy winter skins, worth 2½d. each.

† Black Opossum Rugs are very rarely made from skins in the green state. The prices quoted are for tanned skins in all cases for rugs.

CHAPTER SEVEN — PERSECUTION AND BOUNTY HUNTING

cause of sheep losses in the highlands is theft, with other causes only accounting 1 or 2% of total losses.

Although it was mainly held responsible for ravages among flocks of sheep, the thylacine also had a liking for poultry. But another reason for farmers to make it public enemy n° 1.

Although it is clear that considerable losses in the sheep flocks were attributed to thylacines, many deaths may have been due to other causes, especially kills by dogs. Any animal that dies in the bush may be devoured overnight by devils, therefore leaving no evidence of the cause of death. The east coast landowners claimed enormous losses due to tiger attacks. In an 1886 House of Assembly debate, J. Lyne of Swansea said that 30,000 to 40,000 sheep were killed annually on the east coast, and that a single thylacine would kill up to 100 sheep in a year. He also claimed that the tigers chased sheep over cliffs and that more sheep were maimed than were killed.

According to Lyne, "*these dreadful animals may be seen in hundreds, stealthily sneaking along, seeking whom they may devour, and it is estimated that they will have swallowed up every sheep and bullock in Glamorgan*" as reported in the *Tasmanian Mail* of 3 September. This story is totally incorrect. There is no evidence of pack hunting, and a Tasmanian tiger was never capable of killing a bullock.

In the same House of Assembly debate of 1886, Mr Davies of Fingal declared that 20% of his sheep in mountain country were killed by Tasmanian tigers. Later in the year

TASMANIAN TIGER — A LESSON TO BE LEARNT

CHAPTER SEVEN — PERSECUTION AND BOUNTY HUNTING

Lyne raised his claim to no fewer than 50,000 head of sheep killed every year by thylacines on the east coast, stating that he paid 3 pounds for each thylacine carcass. Unfortunately there is no record of how many carcasses were in fact presented for payment. This practice seems to have been fairly common at the time, and Lyne reported in 1887 that the Malahide estate at Fingal "*paid 25 shillings per head and has killed about 50*".

Sheep losses on the Kelvedon estate near Swansea in the 1885-87 period amounted to about 200 annually, but most were stolen and only a dozen or so lost to Tasmanian tigers in kills made in the hill country towards Tooms Lake. Other landholders in the Swansea area have told similar stories. Mr Shaw kept his losses down to about 6% per annum by employing good shepherds and by avoiding the hill country. This is interesting as it shows that good farming indeed reduced losses, which was also demonstrated on the Van Diemen's Land Company properties. The implication is that high sheep losses were the result of using hill runs, overstocking and poor care of the farm animals.

In 1887 the total sheep population was about 1.5 million. 115,000 of these were farmed on the east coast, which is defined for the purposes of this work as a zone extending from Triabunna to St Helens, inland as far as the watershed to include Tooms Lake. Even allowing for a most generous estimate, the number of sheep in the Swansea area would have been 35,000. The claim that 50,000 sheep were killed by thylacines would constitute nearly 50% of the entire east coast flock, a totally unacceptable situation that would have driven every farmer on the coast to bankruptcy. There is no objection to the claim that thylacine killed sheep, but these claims were grossly exaggerated and losses from other causes were often blamed on Tasmanian tigers.

Left: Every hunter who caught a tiger was proud of the event. Here we have a tiger killed on Penny's Flats near the Arthur River.

In 1832 it was said that "*when its proper food is not available it will attack flocks (and)… is extremely destructive to young lambs, nor do their depredations almost ever extend even to the older portion of the flock*". This sounds more like the predatory activity of the Tasmanian devil who would only kill young or disabled animals. In the same year it was quite emphatically stated that a Tasmanian tiger "*usually kills one sheep at a time*". It is hard to know who was correct, but few other people have reported killing sprees by thylacines.

Contradiction upon contradiction is the regular pattern for Tasmanian tiger accounts, some story-tellers obviously prejudiced and others inclined to make up a better story. It could well be that many simply repeated the gossip of the day, depicting a savage and unrestrained killer slaughtering sheep in a frenzy of blood lust. All this can only be taken as gross exaggeration, and many of the old trappers maintained that the tigers did not act in this way, notwithstanding evidence that in isolated cases, a thylacine did become an addictive killer.

With such an uncommon animal, folklore, superstition and fallacy have inevitably built up around the thylacine. Telling stories around the camp fire at night, shepherds and trappers would certainly embellish their tales, more so if a "townie" happened to be present. From the meagre statistics that are available, it would appear reported sheep killings increased in the 1880 to 1890 period, but this could be as much fact as it could be fiction. The Van Diemen's Land Company diaries do however show an increase in the number of tiger-related incidents from 1887 onwards.

It is important to remember that in the Victorian era, any animal that lived by killing others was regarded as cruel and one that went after domestic stock was to be eradicated at all costs. This also explains the detailed records of thylacine activity on Woolnorth, which suggest that there was an increase in numbers of Tasmanian tigers there, reaching a peak around the turn of the century, which may be characteristic of an increase in numbers throughout Tasmania. Landowners on the east coast began demanding a reward for killing tigers and parliament approved a special rate levied by the Spring Bay Association to destroy these "vermin". In the 1900's the Midlands Tiger Association and Oatlands Tiger Extermination Association were formed, the Hamilton Municipal Council organised a private bounty scheme and the Glamorgan (Swansea) Stock Protection Association paid 2 pounds for every grown animal submitted and ten shillings for every pup. The Government agreed to pay an equal amount to the bounty, provided the claim was submitted through the appropriate association. Hence, a dead tiger was worth 4 pounds, almost the equivalent of a month's wages for a farm hand, although to date no records of any Government payments for this reward have been found.

Right: **Parliamentary Petition for a levy to be raised and used to destroy Tasmanian tigers, 1885.**

Parliamentary lobby groups got busy and a petition signed by 26 east coast residents was presented to Parliament on 28 October 1884, requesting a bounty on Tasmanian tigers. This was unsuccessful and a similar petition was lodged on 24 October 1885, but was also to no avail. The whole matter was again brought before parliament in November 1886, and during the debate it was stated that the farmers should do more to help themselves. In a 12 to 11 vote the government eventually agreed to pay a pound for every adult tiger and 10 shillings for each young animal. No other Tasmanian Parliamentary action has had such a dreadful effect upon the state's fauna. The decision was based on wildly exaggerated claims, which in reality covered up bad farming practice.

There were no attempts made to check the accuracy of the claims, nor was any effort made to ascertain the numbers of thylacines doing the damage. Even a rough calculation based on one kill by a tiger every three days would have shown that a population of over 400 thylacines would have been necessary to kill 50,000 sheep per annum, and there is no evidence from any source that the coastal area supported this number of thylacines in any one year. Nevertheless, with a tiger now worth a month's

1885.

PARLIAMENT OF TASMANIA.

EAGLES AND NATIVE TIGERS:

PETITION FOR POWER TO LEVY RATE FOR DESTRUCTION OF, FROM MUNICIPALITY OF SPRING BAY.

Presented by Mr. Gray, August 14th, and ordered by the House to be printed, August 19th, 1885.

To the Honorable the Speaker and Members of the House of Assembly, in Parliament assembled.

The humble Petition of the undersigned,

RESPECTFULLY SHEWETH:

That your Petitioners are Sheepowners in the Municipality of Spring Bay.—That we have found that the Native Tigers and other destructive animals are making such serious inroads on our flocks that many of us fear we shall have to abandon the Crown Lands occupied by us and give up sheep farming altogether, unless some means can be devised for combating this evil.—That a number of your Petitioners have formed themselves into an Association for the destruction of the Native Tiger and the Eagle, the plan being to pay certain rewards to persons capturing and destroying these vermin.—That the labours of the Association have been crowned by a good measure of success; but contributions being voluntary, the funds have failed, the claims upon them being greater than they could meet, and thus the Association has been compelled regretfully to practically close its labours. And your Petitioners respectfully submit that if these vermin were destroyed from off the Crown Lands, now their chief habitat, not only would present holdings be rendered much more remunerative, but much larger areas of Crown Land would be occupied, to the material benefit of the Colony. And your Petitioners believe that if funds were available for rewarding successful hunters of these pests, much and permanent good would be effected in the cause that your Petitioners have so much at heart, and therefore your Petitioners crave that your Honorable House will be pleased to pass an Act which will confer power on the Municipal Council of this District to levy a rate not exceeding one halfpenny per head on all sheep depasturing in this District, the fund so collected to be employed exclusively for the destruction of the Native Tiger and the Eagle,—or such other measure as will in your wisdom meet the urgency of our need.

And your Petitioners will ever pray.

[*Here follow 26 Signatures.*]

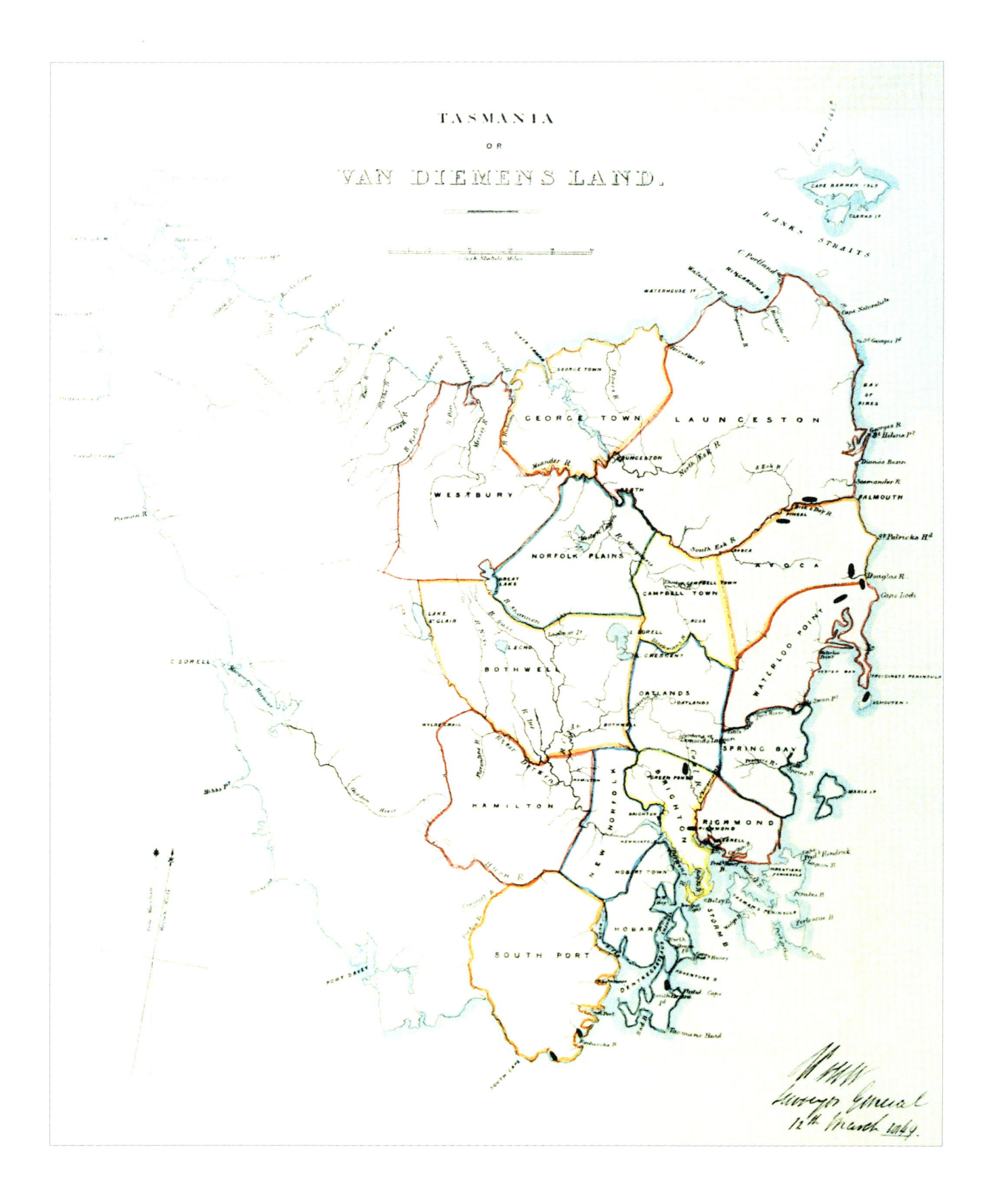

TASMANIAN TIGER — A LESSON TO BE LEARNT

CHAPTER SEVEN — PERSECUTION AND BOUNTY HUNTING

wages, the severe persecution of the species began in earnest. Many men in the country districts, who would not normally have bothered, set out to snare Tasmanian tigers.

On top of this, the Van Diemen's Land Company, not wanting to be out of line, increased the payments under its own bounty scheme to top the government rate. The Company scheme had started in 1830, but with the earliest records lost, only the cash books and some individual accounts show traces of bounty payments. It is known that seventy tigers were killed between 1874 and 1887, which is an average of five per year. If this average were applied to the period of more than half a century following the introduction of the company's bounty scheme, then some 290 tigers would have been killed on Woolnorth station by 1887, which was when the government sponsored bounty scheme was due to commence. However, an oversight meant the funds against which payments were to be drawn were left out of the estimates for the financial year, and in the ensuing debate Lyne again made his claims about the number of sheep losses.

The sum of 500 pounds was provided in the 1888 Estimates, and this amount was allocated through to the 1907-1908 financial year. By this time, the number of claims had declined to such a level that a budget item was no longer necessary. Further bounties were paid by a Governor-in-Council's authority. All of the claims were paid through the Lands Department as illustrated in Table 7.2[2], to which the Van Diemen's Land Company payments at Woolnorth must be added.

Left: An early survey map of Tasmania made in 1849 by R.Power, Surveyor-General of the Colony and successor to G.P. Harris. The district boundaries are shown in heavy lines. There were no roads, few towns and almost half of the State was unexplored and most of the land was only sparsely settled. There was plenty of habitat where thylacines could have lived their lives without seeing a colonist.

To study the biology of a species, a large number of samples are required, and this implies the killing of animals. It is ironic that the government bounty scheme provided such a source of specimens, and yet practically no use was made of material at the time, although the details of bounty payments provide distributional data which otherwise would never have been gathered. A total 2,268 thylacines can be traced to their place of capture through Government, Van Diemen's Land Company or Hamilton Municipality bounty payment records. Some latitude and discretion is required in interpreting the data. For example, there were sixty claims made by residents of Ross and the surrounding district, but it is highly unlikely that these specimens were actually caught in the immediate vicinity of Ross. Discussions with some of the older residents of the town some years ago revealed that the animals were mostly caught in the hills to the east of Ross or in the foothills of the Western Tiers.

Few Tasmanian tigers were caught in the vast tract of mountainous country from Macquarie Harbour to Lake St Clair-Cradle Mountain-Waratah. The thick rain forests that cover much of this country did not support large game populations. The terrain was, and still is rugged and mountainous, not at all suitable for sheep grazing. A few animals

DEPARTMENTAL NOTICE: TASMANIAN TIGER.

From:
« *Tasmanian Forestry, Timber Products & Sawmilling Industry.* »
Dept. of Lands & Surveys, Tasmania, 1910).

DEPARTMENTAL NOTICE

Department of Lands & Surveys, Hobart, 27th April, 1909.

The Governor in Council has been pleased to repeal Regulations referring to the destruction of native tigers, numbered 154 to 160 inclusive, made under the provisions of The Crown Lands Acts, 1903 and 1905. And dated the 30th August, 1906: the repeal to date from and after 26th April, 1909.

By His Excellency's Command,
ALEC. HEAN,
Minister of Lands & Works.

---

may have been caught on the buttongrass plains between the mountain ranges in this desolate country. Figure 7.3[3] shows that thylacines were abundant elsewhere in Tasmania.

The central plateau produced the highest number of bounty claims, with over 660 made in the area. This region was, and still is rich in game and was the scene of extensive winter trapping for the fur industry. It was also used for widespread sheep farming, especially in the summer. Within the central plateau, 233 tigers came from the Dee Bridge-Derwent Bridge district, mostly caught by the Pearce, Jenkins and Stannard families. The fringes of the plateau, particularly on the north-eastern boundaries near Cluan, Mole Creek and Deloraine were highly "productive".

Another large tiger population lived in the Mt Barrow-Ben Lomond highlands and the surrounding woodlands. This large concentration of thylacines extended south-east towards the coast and also in a north-easterly direction towards Gladstone. Apart from the 91 tigers taken in the Bicheno-Cranbrook area, there were relatively few specimens produced for bounty payments in the regions on the east coast, where it was claimed that much tiger predation took place. Swansea only had 17 claims, with 22 in Lisdillon and 23 in Little Swanport. All these animals would certainly have been caught on the hill runs in the ranges between the coast and the Midlands Highway.

Large numbers were caught in the north-west, mostly at Woolnorth and Smithton. The Smithton catch would include all tigers caught in the country districts and collected by the local carter who delivered them to a Police Station and collected the bounty on behalf of the trapper. A substantial number of them probably came from Woolnorth or adjoining properties as well as from further down the west coast as far as Sandy Cape. In bounty times, south-west Tasmania was largely uninhabited, except for men prospecting, felling timber and cutting tracks. In spite of the financial incentive, not many thylacines were taken in the south-west. H. Reynolds, who packed supplies along the Port Davey track, said that tigers were not common, only found on the buttongrass plains and on the coast. He also stated that any thylacines that were caught would probably have been taken to Tyenna for bounty payments. At the time, Tyenna was the entry point to the south-west.

Both the Tasman and Forestier Peninsulas formerly supported tiger populations, with bounty claims being lodged at Nubeena, Bream Creek and Dunalley. Mr T. Dunbabin of Bangor described the Ragged Tier as a particularly tiger-populated area. The geographical spread of the tigers bore no relation to altitude. Thylacines were found throughout the state, and if anything, favoured the coastal plains and scrub. However, open savanna woodland was used extensively and tigers *"were not confined to the mountainous areas"*, as it was frequently stated. This is substantiated by all of the old trappers who talked about Tasmanian tigers, such as H. Pearce, who said that all the tigers caught by his family were taken in the country lying to the east of the King William Saddle, and they never caught any on the other side of the King William Range. These mountains form the beginning of the West Coast Ranges with their associated rain forests, whereas the country in which the Pearce family trapped is open woodland, with plains and abundant game. Catching a tiger was quite an event when they were plentiful and old timers remembered every detail, right down to the very fence where the snares were set to catch the animal.

A successful trapper would take his carcass to the nearest police station, which would then forward the claim to the Lands Department for payment. In many instances the police paid the claim and the Police Department was accordingly reimbursed. A magistrate was also able to issue a certificate and send the claim on to the Lands Department. The toes were clipped off the carcass, or the ears removed so that it could not be presented more than once for a bounty claim. In some instances, the bounty was paid by the Hamilton Municipality through its own scheme and the money recovered through the

Left: **By 1909 there were so few tigers left the bounty provisions were repealed.**
Below: **Distribution of thylacines based on bounty records and sightings.**

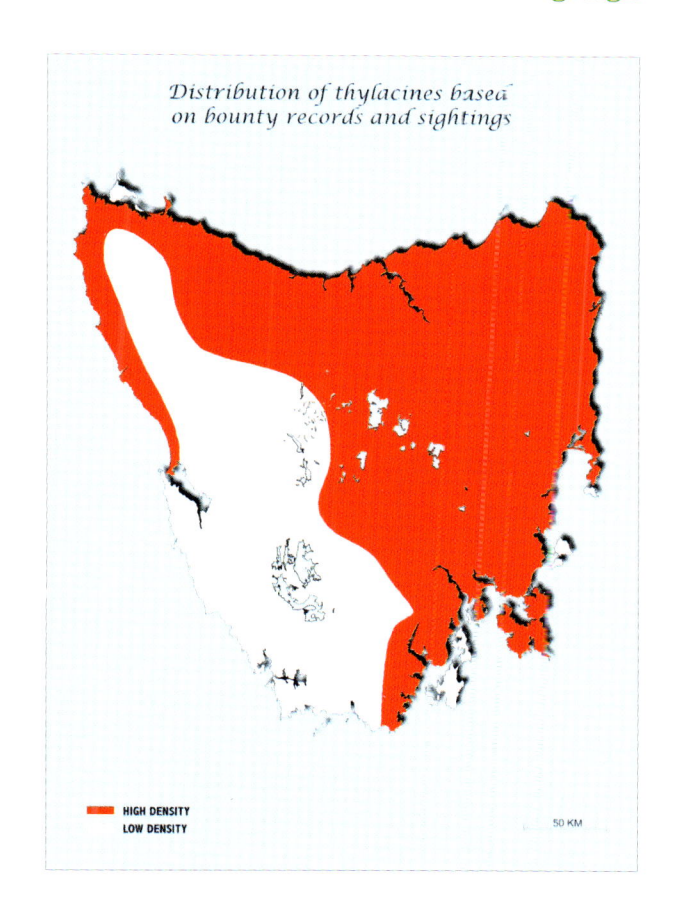

Distribution of thylacines based on bounty records and sightings

HIGH DENSITY
LOW DENSITY
50 KM

TASMANIAN TIGER — A LESSON TO BE LEARNT

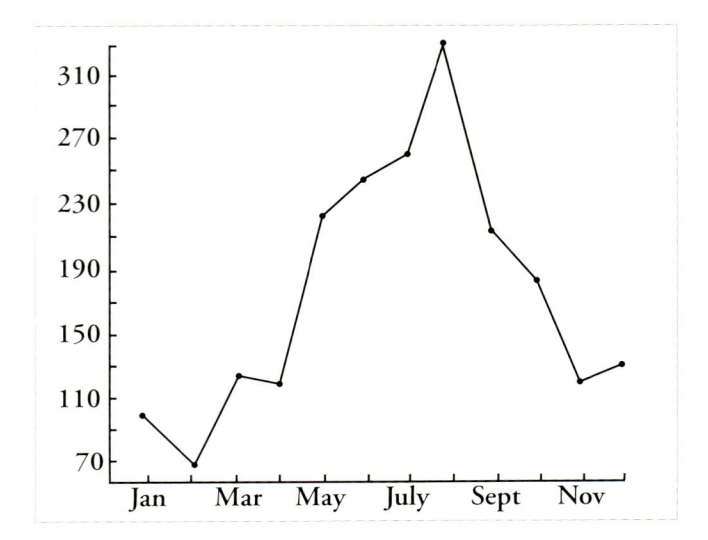

Right: **A time for remorse... This diagram shows that the "peak", in terms of captures, was at the height of the southern winter, as the animal approached inhabited areas, in its quest for sustenance.**

government scheme. The first bounty was paid on 28 April 1888 to J. Harding of Ross, and the last paid to J. Bryant of Hamilton on 5 June 1909. Throughout the bounty period, the amount paid was £1 for an adult and 10 shillings for a pup.

Some basic information has been obtained from searching through the Lands Department account books, which contained the amount paid to the claimant, and whether the payment was for one or more adults or for young animals. The name of the claimant was also usually recorded, as well as the general district in which the claimant was resident. Going through the electoral rolls and the postal directories of the period, it is often possible to determine the property or area in which the tiger had been captured.

It is important to emphasise that the records do not represent the real number of total kills. Many trappers have said that up to half the tigers snared were not submitted for bounty but were carted around and offered to local property owners who paid a reward (usually £1). When the carcass became too smelly it was dumped in the bush. The number of thylacines killed and disposed of in this manner is a complete mystery, and it would be useless to further speculate on the topic.

Opposite: **This photo was taken by a professional photographer, E. Warde of Waratah. He went out of business in 1906 so the photo probably was taken about the turn of the century at a trapper's hut somewhere near Waratah. It is possible that the trapper was J. Cooney who claimed bounty for two adult thylacines in Waratah on 19 June, 1901.**

The Woolnorth diaries give an insight into the intensity of effort that went into the chasing and hunting of tigers. A chase did not necessarily lead to the animal moving away from the area and seldom to its capture. It is hard to say why so much time was spent chasing a thylacine, as the animal would only flee to another part of the property where it would continue to be a

CHAPTER SEVEN — PERSECUTION AND BOUNTY HUNTING

nuisance, but the activity and excitement probably came as a welcome relief from the dull routine and isolation of the farm.

The tiger man operated in the Mt Cameron West area most of the time, and this is where most of the tigers were caught. A number were also taken at Studland Bay as well as on the adjacent Three Sticks. Few were caught on the Harcus or Welcome Heaths, which, in those days, were covered in thick scrub and were not the open paddocks and cleared areas of today. Study of the tables shows that the majority of thylacines were caught on the coastal runs, with only about 15% taken elsewhere. This would be in accordance with the distribution of the tiger's prey which abounds on the consolidated sandhill paddocks adjoining the sea. All 17 tigers caught by G. Wainwright were taken on the coast, mostly in the Mt Cameron cattle yards.

James Malley suggested that Tasmanian tigers moved in a circular pattern up the coast to Woolnorth Point and then along the north coast in an easterly direction as far as Smithton, then south through the bush back to the coast. He may be correct in his assumption, but the evidence from Woolnorth strongly suggests that the tigers were probably resident in one area and stayed on the coastal runs as much as possible.

Tasmanian tigers were caught at Woolnorth during all months of the year, however the highest catch numbers were made in the winter, which is also the pattern in other areas. It has already been mentioned that most trapping took place in the winter months when the pelts fetched premium prices and were less at risk to damage from blowfly strike. The fact that tigers were caught at this time of year is of some biological significance, as it shows that they did not retreat to the hills and rugged areas for breeding, as many of the early accounts suggested.

Table 7.4[4] illustrates sheep losses in the various runs and paddocks of Woolnorth in the 1888-95 period, and there is little correlation between this data and

CHAPTER SEVEN — PERSECUTION AND BOUNTY HUNTING

thylacine catches presented in table 7.5[5]. Once again, this points to the fact that sheep died more from other causes than from thylacine attacks. It may be of little significance, but tigers mostly killed merino sheep in comparison to the other five breeds farmed on the property. It is evident that these tigers were reluctant to move away from their usual area, and this suggests that they established themselves in a home range in the same manner as other carnivores. Hence, the only way to get rid of them was to kill them.

Thylacine catches peaked in the year 1900 at Woolnorth, then declined steadily until only sporadic catches were made after 1903. The 19 kills in 1900 made for a bumper year from which the thylacine population never recovered. In the later days of tiger hunting at Woolnorth, it became clear in the diaries that enthusiasm for the task was not as great as previously, and this could well be due to the lack of successful catches. The snares were left unattended for quite long periods, which implies that it was no longer worth while to go out looking for the occasional animal. The diaries read *"19 May 1905 - Snares set at Mt Cameron"* then only on June 16 *"Mount snares examined"*, and not checked again until July 14. By this time, the thylacine population had decreased to such an extent that

Skins were exhibited in family photographs as at Redpa in 1912.

the tiger man was probably spending much of his time herding and driving cattle, and doing other general farm work.

H. Wainwright made the remarkable claim that G. Wainwright had caught no less than 118 tigers at Studland Bay on the Woolnorth property in 1897, and was apparently paid £1 per head for them. No record of this can be found in the Station diaries, in which everything that happened on the property was reported. There is no evidence of these catches in the Bounty records either, and indeed, 118 kills exceeds the total number taken throughout Tasmania in that year. Around the turn of the century, some landowners near Great Lake established a Tiger Fund of £40, and were prepared to pay 2/6 for every dead tiger. It was alleged that two men caught 300 thylacines and bankrupted the fund in four months. This tale does not stand up to examination, because no trapper would claim only 2/6 when he could get eight times that amount under the bounty scheme. What is more, 300 tigers caught in four months is an unheard of number, and quite unsubstantiated by any other evidence.

Local folk nowadays say that twelve tigers were killed at Native Corners at Campania, but no information has been found regarding these claims. J. M. Dunbabin stated that there was a plague of tigers in the Buckland area in 1870-1890, and that twenty thylacines were killed in the district in that period. It was also claimed that the French family, who lived nearby, killed about seventy. Few of these kills appear to have been submitted for bounty and the numbers may be exaggerated. Mr F. Burbury of Parattah was a treasurer of the Oatlands-Ross Landowners Association, and paid shepherds a bounty of £5.10.0d. for each carcass. In 1953, he recalled that about 40 bounties were paid out - one shepherd collecting on 18 kills. This sum was more than five times the Government bounty and was indeed a large amount of money in those days.

In 1883, Dr Crowther assured that the Tasmanian tiger was not extinct, and wrote that "*hawkers from the interior*" frequently offered skins for sale in Hobart Town. These pelts may well have come from tigers snared by trappers who set gin traps around their possum and wallaby snare lines. Both tigers

CHAPTER SEVEN — PERSECUTION AND BOUNTY HUNTING

and devils were caught in this manner, especially in areas of the central highlands where there was much fur-trapping. The hawkers were possibly responsible for the trade in tiger skins, which sold for £3.18s.0d. when tanned. In 1968 it was reported that 3,482 skins were exported between 1878 and 1896, and were in demand for waistcoats. It was also stated that the skins were exported by an unnamed tannery, long-since defunct, and would not have gone through normal fur-trade channels. No record of this has been found, nor is there any corroborating evidence in the Laird papers in the State Archives.

There is evidence of a thylacine skin being sent as a sample from the Shannon River area to London in September 1861. By the 1880's, there was a market for the skins in Hobart, a good one fetching 10s.6d. (*Mercury*, 26 April 1884). In 1890, skins were quoted at 36 to 72 shillings per dozen undressed, and 48 to 84 shillings dressed, however significantly, the skins were described as scarce.

It is now suspected that dealers and merchants exported many skins without any records being kept, and the authorities had little, if any idea of the large numbers of thylacines actually being killed. All things considered, 3,842 is however a huge number to be taken over a couple decades and therefore this figure cannot be substantiated nor accepted. A small number of animals were caught and exported live to zoos, but records of these are also incomplete. Seven specimens appear to have been sent to London Zoo by Mrs Roberts of Hobart's Beaumaris Zoo and an unknown and probably larger number was sent abroad by J. Harrison of Wynyard.

Left: James Harrison of Wynyard exported tigers overseas during the 1900-1930 period.

Little is known of Harrison's activities in the field. He was a trader of animals and property, and the only records of his business dealings are a 56 page notebook, that was lodged with the Queen Victoria Museum in Launceston in 1996. It is very incomplete, and covers the 1927-28 period. It does show that Harrison sent tigers, as well as other mammals, to several people in Australia.

In 1977, a woman by the name of Doherty told of her father catching a "hyena" in 1916 and Harrison offering £25 for it. This offer was apparently "bettered" by Professor Flynn of the University of Tasmania, who then sold it in Sydney where the dealer could get £300 for it from a zoo. No records have been found of a zoo purchasing a specimen at that time.

In the late 1920's, Harrison bought a Tasmanian tiger from a trapper named Aide Jordan who later in 1987 claimed that this animal was in fact the last tiger to die in Hobart Zoo in 1936. Jordan said that the label on the box containing the animal clearly showed it was dispatched to Hobart, but the minutes and accounts of the zoo show no such transaction. Flynn appears to have had dealings in various species and may have bought and sold both live and dead thylacine. It has been noted that, at times, Flynn was

TASMANIAN TIGER — A LESSON TO BE LEARNT

The tiger, rarer and rarer in the first decade of the 20th century, incited some museums to obtain stuffed specimens.
The animal that is shown in the windows of the Launceston Museum is far from a perfect example of taxidermy. Indeed, the animal looks like a sausage on legs, and this somewhat ridiculous image is a far cry from reality when compared to photos taken of live specimens in captivity.

in trouble with the fauna authorities for some of his activities.

The few catch statistics available from the bounty payments provide some information on the decline of the Tasmanian tiger. In examining the data, some assumptions must be made. Firstly, every snarer hunting kangaroo or wallaby stood an equal chance of catching a tiger. Secondly, tigers were randomly distributed over the country covered by the snarers, some of whom made a special effort to catch tigers. These snarers were readily identifiable and were called professional tiger snarers.

For the purposes of this survey, Tasmania is divided into 6 districts, each of which is easy to identify. The catch from each district has been grouped in five year periods. The data shows that the catch per successful hunter only declined slowly over the hunting period. This can be interpreted as an indication of some degree of over-exploitation of the resource. Although the number of successful hunters declined, the return for each individual remained at a reasonable level.

For the professional tiger hunters, the returns remained high until 1908 when a startling decline occurred in the number of professional tiger hunters, dropping from 181 in 1903-1907 period (36 per annum) to only 16 in 1908. The catch per man dropped from 17 (3.6 per annum) to nil. Such an abrupt decline in the number of successful hunters and catches must be attributed to some catastrophic event rather than to over-hunting, and there is nothing to suggest that these men ceased their activities in 1908. Wainwright continued his snaring efforts at Woolnorth, catching three tigers in 1914.

When the catch is distributed across the six major regions of Tasmania, the return per hunter is lowest on the west coast. This is not surprising, as much of the habitat in this region is not suited to supporting sufficient numbers of large marsupials and maintaining a tiger population in the area. Only the areas around the creeks and along the coastal dunes provided a suitable habitat for these animals, and it is here that sightings of tigers were reported. By far the highest average yield occurred in the Central Highlands and Western Tiers, with over 5 animals per hunter. Clearly this region offered a congenial habitat for Tasmanian tigers, with secluded shelter, plenty of food and reasonably open country for hunting.

It is surprising that the north-west gave low yields despite the continuing productivity of Woolnorth. However, much of the north-west hinterland was occupied by dense rain forest which was unsuitable for tigers or their prey, with thylacines confined mainly to the edges of the forest. The yield on the east coast shows a spectacular high of 13.25 in the 1888-1892 period, but from then on the yield drops more in line with other parts of the State. This collapse is symptomatic of excessive hunting pressure upon the resource, but since the yield later stabilises to around three animals per hunter, it is likely this excessive pressure was not maintained.

Early writers noted that the Tasmanian tiger was rare and lived only in the more remote parts of the State. This observation is so universal that it must be accepted as true. Although data available is most incomplete, it is nonetheless possible to estimate the numbers

A dilapidated hut at the abandoned osmiridium mines at Adamsfield (S.W. Tasmania). This area was the scene of several captures from 1925-34. A female with 3 cubs was captured there by Elias Churchill in 1925 and exhibited around the State. Another animal was caught in 1933.

Places	Date	Sex	Quantity	Notes	Source
Pyengana	1910	F	3	With 2 cubs, sent to Launceston City Park	Graves, 1958
Bermuda	1911	-	1	Shot	Laird Files
Kelvedon	1919	F	4	Three pups	Laird Files
Florentine	1923	-	3	E. Churchill	Laird Files
Trowutta	1927-28	F	3	With 2 pups, J. Home	Laird Files
Trowutta	1928	-	1	Killed by Porteous	Laird Files
Trowutta	1930	M	1	Killed by W. Batty	Guimer, 1985
Rasselas	1933	F (?)	1	Snared by E. Bond Eaten by devils	Laird Files Laird Files
Traveller's Range	1930	-	1	Suffocated in a bag when being carried out.	Q.V. Mus.Files
Mawbanna	1948	-	1	Ray Blizzard. Caught in possum trap. Old and mangy, known to have been in area for years. Clubbed to death.	Q.V. Mus.Files
Pieman area	1949 (?)	-	3		Jordan, 1978
Montumana	1950	-	1	Killed and buried in fowlyard	Laird Files
Deloraine	1952	-	1	Run over by car at «The Huntsman»	T. Hume
Strahan	1961(?)	-	1	Snared	*Mercury* 4 Oct.
?	1960-70	-	6	Malley report	*Mercury* 14/11/70
N.W. Coast	1977	-	1	Shot	Roberts, pers.comm.

CHAPTER SEVEN — PERSECUTION AND BOUNTY HUNTING

of tigers living in Tasmania during the bounty period. Using the Van Diemen's Land Company Diaries as the main source of information, the following assumptions must be made:

1. Each tiger or pair of tigers occupies a home range that is more or less constant in surface area.
2. Each home range is roughly the same in surface area.
3. All of Tasmania was covered by home ranges.
4. Hunting pressure was continuous and was applied as soon as a tiger was sighted.
5. Tigers were not territorial in their behaviour.

The average home range surface area per Tasmanian tiger (or pair of tigers) can be estimated at approximately 55 square kilometres, however this varied greatly from year to year, from a minimum 11 square kilometres in 1900 to a maximum 202 square kilometres in 1879 and 1886. The differences in home range can be taken to represent variations in density of the population because, as seen in the diaries, as soon as a tiger was sighted, efforts were made to catch it. Recognising that hunting pressure was constantly applied to the population, 1879 and 1884-1886 assume special significance, with continuous and immediately applied hunting pressure, it would appear that the population on Woolnorth was effectively wiped out by large kills in 1883, then recovered with migration into the area in 1886. It can therefore be concluded that there was a number of mobile thylacines that chose to settle in the very attractive home range that had thus become available to them.

Left: Table of Tasmanian tigers known to have been killed between 1910 and 1990. Some doubt exists about the reports after 1952, as no carcasses were produced.

It is almost certain that these animals came from the south, since much of the coastal country to the east was settled farm land. Perhaps James Malley was right when he suggested that there was an annual movement of tigers up the coast in the winter. It has been reported earlier that most thylacine hunting occurred in winter (May to October) and the information in Table 7.6[6] suggests that there may have been a winter migration on to the Woolnorth property. Accepting that the home range of a tiger or a pair of tigers was 55 square kilometres, then only four animals could have lived in the area at any one time.

However, in many years, the number of kills was much higher, which implies thylacine movements into the attractive habitat. The Station diaries lend support to the view that there must also have been a resident population because, when a tiger was seen on a particular paddock, persistent hunting would lead to its death. Thylacines were often seen in pairs and sometimes a pair were both taken on the same day, which was the case on 28 January 1901.

THE ANIMALS AND BIRDS' PROTECTION ACT, 1919.
A PROCLAMATION.
I, SIR JAMES O'GRADY, Knight Commander of the Most Distinguished Order of Saint Michael and Saint George, Governor in and over the State of Tasmania and its Dependencies, in the Commonwealth of Australia, in Council, in the exercise of the powers and authority conferred upon me by the Animals and Birds' Protection Act, 1928 (19 Geo. V. No. 51), do, by this my proclamation, transfer Native Tiger (*Thylacinus cynocephalus*) from Schedule 4, Part 1, to Schedule 3, Part 1, of that Act.
Given under my hand, at Hobart, in Tasmania aforesaid, this 28th day of March, 1930.
JAMES O'GRADY, Governor.
By His Excellency's Command.
J. C. McPHEE, for Attorney-General.

Above: It was in 1930 that the Governor of Tasmania took the decision to place the thylacine among those species that were partially protected.
Right: Six years later, one of his successors strengthened this measure by placing the animal in the category of species threatened by extinction, giving it total protection, that, under the circumstances, was all too late.

Some parts of Tasmania did not support many, if any, thylacines. The rain forests and sedgelands, which account for almost half the surface area of the State, were not a favoured tiger habitat. If this area is subtracted from the total area of the State, then the average home range of tigers in Tasmania would be reduced by about one half. The Woolnorth estimates suggest a home range per individual or pair of between 50 and 60 square kilometres, which would indicate a Tasmania-wide thylacine population between 1,357 and 1,138 individuals, or double these two figures (between 2,714 and 2,276) if each home range sheltered a pair of tigers. Home ranges of less than 25 square kilometres would not have provided sufficient space for adequate food, shelter, breeding dens and other elements necessary to thylacine survival.

It would seem that home ranges calculated on the basis of the Woolnorth example are not strictly applicable to the rest of Tasmania. In the Central Highlands, largely trapped by the Pearce, Jenkins and Stannard families, 233 tigers were caught during the bounty period, which gives an annual average of 23 specimens. The area trapped by these families is estimated at some 2,025 square kilometres, which gives an estimated home range of about 88 square kilometres. This in turn shows a total population of 722, or 1,544 if the home range were occupied by a pair of tigers. This figure is much too low to support an average 100 kills a year over a 14-year period.

Therefore, the conclusion is made that there were between 2,000 and 4,000 thylacines living in Tasmania in any one year, and would have been less rather than more. In accepting these population estimates, the annual kills of 109 specimens represents only about 4% or 5% of the total, which would seem sustainable and not have lead to the virtual disappearance of the species by 1908.

In addition to human persecution, thylacines had to endure an alleged epidemic, perhaps pleuropneumonia, which spread through the dasyure population in 1908-1909. This helps explain the rapid decline to almost extinction due to a combination of hunting, habitat alteration and disease acting on an already rare species. The other three dasyures have recovered but the thylacine has not made it.

CHAPTER SEVEN — PERSECUTION AND BOUNTY HUNTING

The bounty period ceased in 1909, at the same time as Tasmanian tigers became rare, and there are no records of catches nor killings during the Great War, after Wainwright's capture of 4 animals in 1914. But in spite of this rarity, the thylacine was afforded no legal protection. Classed as unprotected, this meant anybody could catch, sell or kill a tiger. In comparing the small numbers of tigers caught in the 1910-1990 period to numbers from the bounty period, the serious and rapid decline of the tiger population becomes obvious. Indeed, the 30 animals caught since 1910 represent about one-third of the annual kill during the bounty years.

Although it must have been apparent to all by about 1920 that the Tasmanian tiger was a very rare and highly endangered species, nothing was done. This was not only due to the apathy of the Tasmanian people, but also to the structure of the fauna administration of the day. The control of fauna was the responsibility of the Police Department, hence the Commissioner was the Chief Guardian of Fauna. There was no independent body supervising or advising the Government on related matters, and decisions were made by police acting on advice from the local constable. The majority of decisions involved the open seasons on fur-bearing mammals. This changed in 1929 when the Animals and Birds Protection Board was established to administer laws relating to fauna, with the authority to enforce its decisions.

> THE ANIMALS AND BIRDS' PROTECTION ACT, 1928.
> A PROCLAMATION.
> I, SIR ERNEST CLARK, Knight Commander of the Most Honourable Order of the Bath, Commander of the Most Excellent Order of the British Empire, Governor in and over the State of Tasmania and its Dependencies, the Commonwealth of Australia, acting with the advice of the Executive Council, in exercise of the powers and authority conferred upon me by the Animals and Birds' Protection Act, 1928, do, by this my proclamation, transfer Native Tiger *(Thylacinus cynocephalus)* from Schedule 3, Part 1, to Schedule 2, Part 1, of that Act.
> Given under my hand, at Hobart, in Tasmania, this 10th day of July, 1936.
> E. CLARK, Governor.
> By His Excellency's Command,
> E. J. OGILVIE, Attorney-General.

Meanwhile, throughout the 1920's, the Tasmanian tiger was liable to be shot, snared or persecuted, and no public body or vigilante organisation appears to have brought pressure to bear on Government to grant some form of protection to the species. In fact, attempts to catch tigers for zoos went on well into the 1930's. It wasn't until 20 August 1929 that a motion was passed by the Animals and Birds Protection Board providing for the partial protection of the Tasmanian tiger. The Board closed the open season for the month of December, a regulation that was only gazetted in May 1930, which meant tiger hunting was only prohibited in December that year. The rationale behind the decision was that December was believed to be the month in which breeding took place.

With the benefit of hindsight, it is difficult to imagine what the Board actually tried to achieve by the "partial protection" measure. Virtually no-one was hunting tigers and those that hoped to catch one for a zoo were usually unable to obtain a specimen. Clearly the animal was in a perilous situation, and protection for one month a year was

> **BAY VIEW HOTEL, STRAHAN, 1975**
> **MENU**
> Meal Hours   1.15 a.m. — 2.30 a.m.
>
> | | |
> |---|---|
> | Consomme of Blue Tongue | $0.50 |
> | Parrot Pie | $1.50 |
> | Kentucky Fried Ferret | $2.50 |
> | Crumbed Coot | $2.00 |
> | Sweet 'n Sour Devil | $4.50 |
> | Curried Cormorant | $0.50 |
> | Fur Burgers | $5.80 |
> | Tasty Tassie Tiger | $2420.50 |
> | Goanna Grits | $0.80 |

This document is the record of the rarity of the species as displayed in the bar by a humorous landlord. The price for a "Tasty Tassy Tiger" has skyrocketed since then!

totally inadequate. Eventually, in September 1933, a sub-committee was appointed to consider protection measures for the species, but little appears to have come from its deliberations. Indeed, permits to catch tigers for zoos continued to be issued up until 1936. When partial protection failed to produce the desired results, concern for the future of the species was expressed at several Board meetings, and the Tasmanian tiger was finally granted full protection on 14 July 1936, a status still enjoyed today...

It is somewhat ironic that by the time the thylacine was granted full protection, the first of the major searches was being contemplated, in the form of an expedition organised by the Board, which belatedly departed in November 1937. The administrative action was far in arrears of the needs of the species and should have been taken in the early 1920's, when it was already apparent that Tasmanian tigers were becoming rare and were in need of full protection. In some sectors of the community, the attitude adopted was one of outright antagonism, a general apathy that is difficult to comprehend in modern times. On 2 February 1965, the Government of the United Kingdom banned the import of Tasmanian tigers, but it was not until 26 June 1973, and much too late, that the Australian Government woke up to the situation and introduced legislation banning all thylacine exports.

When the Gazette Notice granting total protection appeared in 1936, the suggestion was made to round up the surviving Tasmanian tigers and move them to De Witt Island, which was, and still is, a miserable place off the south coast with no suitable habitat and very little prey for a thylacine. No suggestion was made as to how the animals were to be rounded up. Fortunately no tigers were in fact herded, sparing the species from further persecution.

Various approaches were made to government at different times, suggesting that attempts be made to capture thylacines and then rehabilitate the species in some sort of sanctuary. These suggestions started when Professor T.T. Flynn, then the Ralston Professor of Biology at the University of Tasmania, foresaw the future scarcity of the species and proposed that some specimens be captured and relocated to an island. In the late 1930's, Summers and Sharland, both involved in early searches for a thylacine, respectively recommended that sanctuary areas be established between the Arthur and Pieman Rivers

CHAPTER SEVEN — PERSECUTION AND BOUNTY HUNTING

and in the Jane River area. Unfortunately the suggestions were not taken up at the time and to this day there is no reserve in these regions.

A sanctuary was not easy to establish, and would have had to support enough animals in a habitat suitable for both prey and predator. It would also have had to be of sufficient surface area to allow movement for the animals, provide resting sites and still prevent "overflow" of the predators and prey into neighbouring properties, which was an altogether difficult arrangement to achieve. The larger State parks certainly offer sufficient cover and food for the tigers to flourish. Places such as South-West Park and the large Cradle Mountain-Lake St. Clair Parks appear to offer the habitat in which thylacine could live and breed, however these parks are not in areas that previously sustained significant populations of Tasmanian tigers, which inevitably leads to the conclusion that the habitat was not attractive to the species.

The notion of a reserve for tigers does not seem to have been pursued after the Summers and Sharland suggestions. Perhaps the animal had become so rare, if not extinct, that the idea was not worth pushing. It was not until the 1963 Board expedition to catch a Tasmanian tiger that the question of a sanctuary came to the fore, largely in the context of what to do with a tiger if one was caught.

All of the existing parks were rejected and the Board started to search elsewhere. The concept widened from a tiger reserve to a nature reserve where the public could see native animals in their natural environment. The Board eventually settled on Maria Island, off the east coast. The island offers a good thylacine habitat, plenty of prey species with the added advantages of cheap land, no fencing requirements, remoteness from poachers and the various species impounded there could not invade surrounding properties and become a nuisance to landowners.

The reserve was duly acquired and is now established as an animal viewing park. There is an aura of sadness about it, however, because there are no Tasmanian tigers to put there…

1- See appendix table p.230
2- See appendix table p.231
3- See appendix table p.231
4- See appendix table p.232
5- See appendix table p.233
6- See appendix table p.233

Chapter Eight
Tales of

the tiger

*O*ver the years, many tales have been told by men who actually caught Tasmanian tigers. It must be remembered that 1908 was the last year of real thylacine abundance, and that the men who caught tigers for bounty and knew about their habits have all left this world, and tales told today are, at best, second-hand. The few captures made in the 1910-1930 period are still vivid memories for those involved, like Wilf Batty who remembers every detail of the incident when he shot a thylacine at Mawbanna.

The stories that follow were gleaned from old timers around the 1950's. Many of these men were active both mentally and physically and recalled incidents from earlier days in the bush. Nearly all of them had figured in the Lands Department bounty accounts. The majority of them paid the price for a rough existence in wet conditions, suffering from rheumatism, arthritis and other ailments in later life. But their eyes would still gleam when speaking of the *"hard but good old days in the bush"*, and almost to a man, would have continued to search for the Tasmanian tiger had it not been for the stiffness in their joints.

Today, the Central Highlands are readily accessible by car, but the area was not opened up until the Lyell Highway to the west coast was built in 1923. Before that, there were only a few rough cart tracks into the area and the region was isolated, even from the small town of Ouse. Living in solitude in the Black Bobs-Strickland-Derwent Bridge district for many years made the Pearce family a reticent, self-contained unit, not given to talking to strangers or officials from the outside world. Two members of the family have talked about their experiences with Tasmanian tigers. One of H. Pearce's first questions was: *"Why did you want to go and protect them bloody useless things?"* He never liked

Tasmanian tigers, although he made quite a lot of money from them, claiming bounty on 53 specimens, whilst his relatives claimed on a further 23.

Pearce's view was much the same as that universally adopted in various degrees of intensity by the old-timers, particularly those who were shepherds or owned sheep, that thylacines should be eliminated because they killed sheep and lambs and were *"of no use in the bush"*. There was no argument about it and they had absolutely no regrets about their deeds, nor is there any reason why they should have, for they knew nothing of the value of a predator in a natural ecosystem and the present-day outlook came too late to reach them and hence save the Tasmanian tiger.

A few did express sorrow that thylacines had vanished from the bush. In a conversation some time in the early fifties, H. Pearce said: *"I put up a slut and three pups out of a patch of man ferns about five years ago"*. The area where this incident occurred is now at the bottom of Lake King William and Pearce said he turned his dogs on them, but continually dodged the issue as to whether the tigers were killed or not. It is likely that they were, and he knew that it was wrong to kill them.

Previous page:

"Encounter with a Tasmanian tiger", from the illustrated Sydney News, April 1867.

Allen Briggs of Safety Cove has kindly provided details of an interview he had with J.M. Dunbabin of Dunalley on 8 September 1961, who reminisced as follows:

"Tigers were a plague in the Cockle Bay area around 1870-90 when they killed sheep. Lots of hunting and snaring. Poisoning was no good as they never came back to a kill a second time. We caught about twenty all told but got no relief until we bought land on Forestier Peninsula and moved all the sheep. They apparently live in the Sandspits-Pony Bottom area. French's at Buckland got about seventy. No sheep were killed after about 1890-95 on our place. The usual way was to set snares in the fences. At Cockle Bay they had a yard with decoy sheep in it where they expected the tigers to come. They usually started to chase sheep on Bedding hill and killed when they reached the level. Not having speed he is supposed to wear his victim out with a steady chase".

The numbers he quotes are disturbingly high. In all, the Dunbabins and the French's presented only four animals for bounty, and even allowing for some exaggeration, this is still a long way short of the totals mentioned above. Probably the farmers were only too pleased to get rid of the pests and did not bother much with the bounty, or alternatively the skins could have been offered through the trade for manufacture into waistcoats.

Discrepancies appeared elsewhere in interviews, and it is clear that many more Tasmanian tigers were killed than appear in the bounty accounts. There is no doubt that tigers lived in the Sandspits-Pony Bottom area as most of the country was underdeveloped

CHAPTER EIGHT — TIGER TALES

at that time. Nowadays, areas of it are being logged for woodchip and the habitat has been ruined. Dunbabin himself said that the Ragged Tier near Dunalley produced a few tigers. He added that *"tigers were the most cowardly animal in the bush and a strand of packing twine would hold him"*. This is but another account of an animal which gives up easily and is quiet and easy to handle once caught. His other remark that tigers never returned to the site of a kill is in line with the opinion of other old-timers.

R. Stevenson was a snarer operating on Aplice near Blessington, and he laboriously built about 3 kilometres of wire netting fence to guide tigers into pitfall traps with swinging lids which he had dug to about two metres deep. The lids tipped the animal into the hole, and, as quoted by S. J. Smith in 1980, Stevenson allegedly caught sixty in this way between 1890 and 1906. The method used by Stevenson is interesting because it shows that it was well worth while for a trapper to exert considerable effort to catch thylacine, despite further discrepancies between the bounty records and the catch numbers claimed.

Dunbabin's statement that sheep were chased on the Bedding hill is interesting. In

The Pearce home at the Clarence River, Central Highlands, February 1984. This family caught many tigers.

1927, Willett, a trapper operating on the Island, put forward the same view when he stated that sheep were attacked at night while asleep. The slow persistent chase by Tasmanian tigers is also frequently reported.

Mrs Louisa Ann Meredith, a very active woman, arrived in Tasmania in October 1840, showing great interest for her new country and its natural history. She collected and painted seaweeds and plants, writing of her experiences in the new land. Her observations are both perceptive and reliable, it was she who sent a tiger to the Wilmot Zoo at Government House, which was the first live thylacine to be exhibited. The specimen was a cub whose mother had been killed by a shepherd. Mrs Meredith wrote:

"No care or kindness will civilise it even when taken young. Not heard of very often now. A skin was 4'6" from head to tip of tail. Enduring but not a swift runner proceeding at a canter. A party of bushmen saw a kangaroo hopping along followed about ten minutes later by a female tiger scenting along the track and a quarter of an hour later two cubs came along the same track".

Meredith's first statement is controversial. In 1851, Gunn had stated that his captive was *"becoming tamer and it seems far from being a vicious animal at its worst and the name tiger or hyena gives a most unjust idea of its fierceness"*. There is a fair amount of verbal evidence that those tigers that survived post-capture trauma and were held in captivity became quiet, but not domesticated. However, it is hard to believe stories that tigers were held on a chain and taken for a walk like a dog (*7ZR* radio broadcast, 16 February, 1994). There is a tale of a tiger that was held in a cage in a kitchen prior to being sold, and of another tiger that was allowed to roam around a house. One of the tale's describes the animal as being held on a chain, but it bit the glass off a lamp and all present were terrified. One cannot help but wonder how the residents tolerated the smell.

Most marsupials make poor household pets, they are difficult to train, some of them smell more than a bit, and the larger ones tend to take up a lot of space and are not noted for endearing, intelligent behaviour. The description of the hunting procedure is very interesting as it suggests that the sense of smell is important, perhaps more so than is realised by those who have said that tigers hunted mainly by sight.

The use of olfaction would indeed make night hunting much more effective, and Meredith's description includes the cubs using smell to follow the mother. Clearly they were too large for the pouch and had to tag along as best they could while the mother hunted. This supports H. Pearce's view that they did not have lairs and that the young followed the mother when small, and were vulnerable to attack by Tasmanian devils.

The following description of the wreck of the *Acacia* at Mainwaring Inlet on the south-west coast appeared in the Hobart Mercury on 23 May 1905:

Left:
"One night when I had number of possums on my back a tiger followed me..."

"*The bodies had been gnawed by Tasmanian tigers, one of which was found dead near the corpses, apparently having been bitten by snakes, many of which were found near the bodies and killed by the party*".

Tigers have never been accused of necrophagy and it is likely that this was the work of devils, a species that will eat anything. Presumably, it was a thylacine corpse lying nearby, as tigers had been reported from time to time in the Mainwaring River area.

George Johns of Deloraine, although too young to be on thylacine bounty lists, spent his lifetime snaring in the Western Tiers and Arthur's Lake areas as his father had done, and he said:

"*Tigers won't go over anything when they can go under it. You can set deadhead snares and they won't even strangle themselves*".

The first observation is very interesting and should be noted when looking for footprints. The second relates to the docility of the tiger in a snare. There are stories of wide areas around a snare being damaged and the animal having escaped, but this was

TASMANIAN TIGER — A LESSON TO BE LEARNT

more likely to be the work of devils, which certainly do tear up everything within reach. There is also the tale about a tiger making a big leap and breaking the snare. S. Woods of Ringarooma tells of this incident which occurred in the foothills of Mount Victoria:

"In 1934-35 when I was about 17 years of age a tiger went round my snares eating the possums. One night when I had a number of possum on my back a tiger followed me. I could see his eyes in the light of the torch but he followed me all the way home. I was scared".

Woods had seen this animal a few days previously and was able to give a good description of it. There would seem to be no doubt that it was a Tasmanian tiger. The habit of following humans has already been noted on several other occasions. W. Sawford of Parattah, who died on 7 August 1977 at the age of 99, was probably the last tiger man in the midlands, collecting bounty on nine specimens. He was interviewed in 1960:

"There weren't a lot about but a few could be a terrible nuisance to sheep, scaring more than they killed and scattering the mobs. I caught most of mine in the Swanston area towards the coast".

"The female sprang at him, the pistol wouldn't fire…"

CHAPTER EIGHT — TIGER TALES

All the literature clearly states that there never were many thylacines, and it is important to see this supported by a bushman. Sawford's comment that more sheep were scared than were killed is supported by statements from others who contended that good shepherding reduced losses to tigers to a minimum.

C. Thomas of Oatlands has an old diary from the 1868-1870 period and there is no mention of "tigers" in it. Thomas said that he caught most of his tigers east of Oatlands between Lemont and Tooms Lake and in the Ross area. The last one he killed was at St. Peter's Pass. The trapping effort by Thomas, Willett and others was mainly to the east of the Midlands. D.D. Davis of Ross remembered that a few thylacines used to live on Badajoz Tier near Lake Leake, but he could not recall any being seen there since 1900. Here again, the emphasis appears to be on hunting in the Eastern Tiers.

Snarers worked in the Interlaken region and the bounty records show twelve tigers taken there. A little further south near Bothwell at The Den, there was a hut which had a number of thylacine skulls nailed on the wall. The hut collapsed many years ago and the skulls disappeared.

The following comments were made by three pastoralists on the east coast:

"Kelvedon losses in the 1880's were about 200 sheep per annum, mostly stolen but about 12 per annum to tigers, mainly south of Swansea and inland towards Lake Tooms. Not many were lost on Mayfield" (T. Cotton, Kelvedon).

"Lisdillon and Cranbrook had about 2,000 sheep but about 10-12 per annum were mauled. Not many tigers, only about six a year were claimed as a bounty but we didn't use the back country on account of tigers, but didn't lose many sheep on account of good shepherds" (F. Shaw, Red Banks).

"Lost about 150-200 sheep per annum in the 1880's but all were stolen. No thylacines about" (R. Amos Cranbrook).

So much for the arguments of the pro-bounty lobbyists in parliament. The three gentlemen quoted above farmed in an area from which many bounty claims were made, Bicheno-Cranbrook producing 91 animals. Yet sheep losses were comparatively light, as a result of good care of the flocks. These farmers suffered more losses from thieving than from Tasmanian tiger predation, as did the Van Diemen's Land Company. Shaw's claim of six thylacines per annum presented for bounty seems excessively high.

In 1928, the director of the Tasmanian Museum, C. E. Lord, wrote: *"Tigers cannot be held safely by the tail"*. Other writers and many old-timers stated the opposite. It is possible to immobilise many species by holding them by the tail, and the rigidity of the Tasmanian tiger's tail would enable this to be done effectively and with a certain amount of security to the holder.

In 1939, a trapper was quoted as saying that tigers could not swim. This is highly unlikely because swimming is something most mammals can do. Tasmanian Aborigines described the thylacine as a very strong swimmer, swimming like a dog with only the top of the head and ears above water, using the tail as a rudder. The tail is slightly laterally compressed and this led to the erroneous suggestion in 1846 that the species was aquatic and piscivorous.

Right: Photo of T. B. Moore taken in mid-1880 when he was 29 years old. The tiger skin cap was made from the hide of an animal killed by Moore's dogs earlier in the year. The dogs' names were Spero and Spiro (on right). Moore named the Spero River and Spero Bay after one of the dogs, and the other is commemorated in a mountain range. All of these locations are on the south west coast.

"Thylacines only eat freshly killed food" - the statement appears time and again in the literature and in the wild this is undoubtedly correct. Trappers agree on this point, indeed it is one of the points on which there is general consensus. In captivity, Tasmanian tigers were known to eat all sorts of dead material, which could suggest they would do the same in the wild, however there is no evidence of any such habit. In 1960, the unusual comment was made that Tasmanian tigers raided garbage dumps in search of fat, and bread in particular. This view has not been supported by others, but there is no doubt that tigers did hang around bush camps, but whether this was to glean scraps or out of simple curiosity remains debatable.

In the *Weekly Courrier* in 1924, a correspondent's account went as follows:

"A bushman carrying an automatic pistol for protection against tigers encountered a large male and young female tiger near Waratah. The female sprang at him, the pistol wouldn't fire so he hit it with some wood. The male leapt at him and he hit it with a stick and it crawled away whimpering. Later, the man saw a cub and thought this was the reason for the attack".

This tale gives the Tasmanian tiger a more fearsome nature than trappers ever found it to have, although the female may show aggression to defend her young. Years later, the bushman's brother, C. Penny, recounted the incident. The bushman had startled two young tigers from their hide in the scrub and the mother chased him. In defence, he hit her on the head with a piece of timber which had a nail protruding from it and this penetrated the back of the tiger's neck and killed it. This shows how an incident can be changed to make it a "good story".

The next tale is taken from a letter submitted from Forth. "J.S." is believed to have been James 'Philosopher' Smith who lived at Forth at that time. Smith explored and prospected the hinterland, eventually discovering the Mount Bischoff tin field which was the richest mine in Tasmania during its period of operation.

"Two tigers seen on Black Bluff in November 1862 were attacked by a setter dog which seized the tail of a tiger and was dragged off. They fought and the dog throttled the tiger. Later on another large tiger was seen but it ignored the dog. The man tried to hit the tiger with a tomahawk but it ran away and fought with the dog, both biting each other. The man hit the tiger on the head with the tomahawk but it continued to fight the dog for two minutes. The man seized its hind legs and severed the neck with the tomahawk. It was a female with four pouch young, 21 inches high, and 3 feet 4 inches nose to rump length, and 17 inches tail length. Tigers never attack a man but will follow them for considerable distances at night but not molesting him. They retreat to a den when attacked. Sheep dogs and tigers are equally matched but the former wins the fights but is severely injured in doing so whereas kangaroo dogs win easily".

This story has a ring of truth about it and sounds much more authentic than the pistol episode. It is noteworthy that the tigers did not attempt to attack the man, not even when the unfortunate animal was being walloped with the tomahawk. This female was a large animal, being well up in the size range for the species. The statement that they will follow a man at night has been encountered before - in Wood's experience - and in daylight too according to others. Retreating to a den would be the habit of any animal not able to run away.

The exploration of south-west Tasmania was largely associated with prospecting for minerals and timber. All of the region of about two million acres west and south of a line from Strahan to Queenstown to Ouse was untouched by roads or even tracks and only accessible to the intrepid few who walked and cut tracks for themselves and their pack-horses. One such individual was T. B. Moore who explored south from Macquarie Harbour in the 1880's, naming various features of the land, including Moore's Valley. A fascinating photograph was published by Binks in 1980, showing bushman Moore in a posed studio photograph with his two dogs Spiro and Spero. Moore was wearing a "Davy Crockett" hat made from the skin of a thylacine that had been killed by his dogs. The last authentic killing of a thylacine in which a carcass was produced was the now well-known incident at Mawbanna, when Wilf Batty shot a large male in 1930. The story, as related by Wilf, is very similar to the following narratives:

"The animal had been in the area for some time and had been seen by some farm workers in the district. They were frightened of it and wouldn't go near it. Wilf was in his house and heard a scuffle and when he went outside with his gun he saw a tiger with its head under the wire mesh around the henhouse. The animal moved off and went clockwise around a nearby shed. Wilf went anti-clockwise and they met at the far corner of the shed whereupon they both reversed directions and met again at the opposite diagonal corner of the shed, nearest the henhouse. At no time during these manoeuvres did the tiger hurry or seem frightened. Wilf attempted to catch it by the tail but couldn't as he was carrying his gun in one hand. The thylacine reversed direction again and headed for the fence whereupon Wilf shot it. Both his kelpie dogs were terrified by the

CHAPTER EIGHT — TIGER TALES

presence of the dead tiger and did not go near the house for three days. During the time the tiger had been in the area it was seen to jump a 4 feet 9 inch wire fence, touching only the top strand with its feet".

The total lack of fear shown by the tiger is typical and has been reported often. That the animal had been around for some time supports the temporary home range theory, and this is another instance of an animal hanging around outhouses either out of curiosity or more likely hoping to get a tasty chicken dinner.

Another story written in 1910 comes from north-east Tasmania:

"At the time my father was connected with mining at South Mount Cameron and a popular outing was to climb Cube Rock on the Cameron Range. We always sighted tigers or hyenas as we commonly called them and they were much larger than reports I have read about. They were as big as an Alsatian dog and would not move out of your way. We always went around them and always carried a rifle, though they never attacked".

Quite a credible although strange tale, that the thylacine did not move off the track and the party had to go round it, but this further illustrates the absence of fear of humans.

TASMANIAN TIGER — A LESSON TO BE LEARNT

The relationship between dogs and Tasmanian tigers is yet another area where there is contradiction in the records. The most widespread opinion is that dogs showed considerable fear in the presence or vicinity of thylacine and would seek shelter and protection from their owner, or run away and hide. This view was held by most of the trappers.

An incident near the Walls of Jerusalem in 1958, reported three years later, is typical of these stories: a cattle drover was camped in a log cabin with his three dogs…

"*After dark following a scuffle in the bush, two of the dogs came into the cabin. Next day he found a third dog dead with its heart eaten out. He took his horse and two dogs to a nearby gully and the dogs ran under the horse and shortly later a tiger appeared on a rock*".

Horses were also known to be reluctant to proceed further, sensing the presence of some strange animal and laying back their ears. H. Pearce said that tigers could easily be killed by cattle dogs and kangaroo dogs but that the dogs had to be trained to do so. He said that smaller sheep dogs would find the going hard and breeds of small dogs would lose the battle. A fight between a Tasmanian tiger and a bull terrier was described in 1927, when the tiger came out the victor by biting off the top of the dog's skull and leaving the brain protruding.

Previous page: "Next day, he found a third dog dead…"
Right:
"He took his horse and two dogs to a nearby gully…"

A story from what is now Campania House, in the *Hobart Town Gazette* in 1823, presents a different picture from the early days of settlement:

"*A few nights ago, a hyena, an animal so rarely seen in the Colony, was found in the sheep fold of G.W. Gunning, Esq. J.P. at Coal River. Four kangaroo dogs, which were thrown in upon him, refused to fight and he had seized a lamb, when a small brown terrier of the Scotch breed was put in and instantly seized the animal, and after a severe fight, to the astonishment of everyone present, the terrier succeeded in killing its adversary*".

This tale is of note because it suggests that, even in 1823, not many tigers had been seen in the Colony and gives support to the view that they were not common at the beginning of colonial settlement. It also supports Pearce's view that even kangaroo dogs had to be trained to attack Tasmanian tigers.

The general picture of the thylacine is of a docile creature which does not attack people, even under provocation, and this view is held by most of the trappers interviewed by the author. In 1939, the thylacine was described as a timid creature, telling of one specimen which was surprised inside a hut near Adamsfield, but made no attempt to attack the man who entered through the doorway, and was only anxious to escape. Sharland also recounts a trapper's experience when he and a companion found a

CHAPTER EIGHT — TIGER TALES

TASMANIAN TIGER — A LESSON TO BE LEARNT

Tasmanian tiger in a snare. The trapper released the animal which made no attempt to attack him but contented itself with glaring at the companion who by this time was up a tree.

It has been mentioned how the rigidity of the tiger's tail enabled the animal to be effectively held by the tail, and this feature was used to advantage by Mrs Allen when handling one snared thylacine. The narrative comes from Hedley Allen in 1958, describing events in the Seymour district of the east coast:

"My mother… around the snare line discovered a snared tiger… she seized it by the tail at the moment the snare broke. She was able to avoid the snapping jaws of the enraged animal by

CHAPTER EIGHT — TIGER TALES

jerking its tail in the other direction as it tried to bite until her brother laid it out with a stout waddy".

There are accounts of people being bitten by a captive thylacine, even zoo owner David Fleay suffered this attention in Beaumaris Zoo. There is no doubt that a captive or perhaps cornered animal would bite if given the opportunity, or if its young were threatened. Mrs Roberts used to move among the tigers in Beaumaris Zoo and was never harmed.

Another tale from 1939 depicts a trapper on his way back to Fitzgerald when he was overtaken by a storm, finding the door of his hut open when he got there. He entered and found a Tasmanian tiger inside, that immediately dashed out past him into the snow. The trapper later heard something sniffing at the door and when he opened it there was the tiger that again bolted away. The trapper fired both barrels of his shotgun but missed.

Left: Milton grabbed it by the tail...

Hobart naturalist M.S.R. Sharland narrated another humorous tale of a tiger held by one leg in a snare. The snarer was accompanied by a "*new chum*" who climbed a sapling to get out of the way of the released tiger which circled the tree. The fellow tried to climb further up the sapling which started to bend over, causing the climber to panic and yell "*shoot it*". The trapper thought it was all too funny to pull the trigger and the tiger loped off.

A very early story tells of G.A. Robinson at Sandy Cape in 1834 with a party of Aborigines, accompanied by dogs, that surprised a female thylacine and her three pups as they fed on a kangaroo. The mother ran away over the plain followed by a dog that bit her on the rump. The tiger then turned around and chased the dog, and the men rushed to help but the tiger turned on them and the dog killed the tiger. Two of the cubs escaped, and the female's skin was saved.

There is an undated story, probably from the 1920's, that a tiger was caught by F. Milton at Milabeena in north-west Tasmania. Milton shut it up for the weekend in a hut, and when he went back in the animal tried to escape up the chimney. Milton grabbed it by the tail and put it in a chaff bag, later displaying the animal at the Wynyard pub, before eventually selling it to Harrison.

Some of the tales sound rather far-fetched and probably have been enhanced in the telling over the years, such as the claim in the late 1800's that no less than 17 adults and some pups were killed by three men with dogs in a single day. This party set out near Moulting Lagoon on the east coast and *"killed four hyenas in the scrub"*. More were shot in nearby bush and some others took to the water and were shot while swimming. The pups, who wouldn't swim, were also shot.

This tale was retold to W.C. Cotton (*Mercury*, 6 March, 1971) by "old-timers", but all of this seems most unlikely as it suggests a density of tiger population far greater than was ever known elsewhere. It also implies that there was a pack or even packs of tigers in the area. Many experienced bushmen have said that there was no pack behaviour and Cotton himself who lived on the east coast saw only one individual. The animal, a female with pups, was killed and the pups were given to Cotton as pets but they were killed by his dogs.

CHAPTER EIGHT — TIGER TALES

An amazing tale in the Saturday evening *Mercury* of 7 November, 1973, is that of Fisher, in which she claims a tiger jumped on the running board of her car at Hellyer Gorge and barked at the driver… but tigers don't bark…

In 1953, James Lefevre related his reactions to an encounter *"with a big bull tiger that showed fight when suddenly I came upon him. This fellow had just made a kill and his face was covered in blood. When I say a big one this fellow was about the size of a full grown Alsatian dog and being alone in the bush at the time without a knife or any other weapon of defence, I took the line of least resistance and made for my camp"*.

A tiger shot at Fitzgerald in 1912 had bitten its killer on the foot and a man was bitten on the leg when he tried to separate a dog and a thylacine that were fighting at Thompson's Marshes near Chain of Lagoons on the east coast. George Stevenson relates that he caught a tiger and was carrying it home on horseback in a sack. The tiger managed to turn in the sack and obtain enough grip with its paws on the horse's rump to enable it to bite his shoulder.

An account in the thylacine file of the State Archives tells of a Miss Murray who was doing her washing in a west coast creek when she was attacked outside her house by a

Much water has flowed under the bridge since this day in the year 1930 when Wilf Batty proudly posed next to the tiger he had just shot at Trowutta (*left*). This smiling elderly gentleman in his twilight years is being interviewed by New Zealand journalist Elaine Murphy, in 1989 (*above*). Gone forever, the legendary animal, just like the wonderful years of youth of the former hunter.

159

TASMANIAN TIGER — A LESSON TO BE LEARNT

one-eyed thylacine that bit her severely on the arm but then fled after she trod on its tail. The scars could be seen on her arm when she was in her eighties. There is also a story about a west coast resident who killed a Tasmanian tiger after it chased her into her hut whereupon she closed its neck in the door and belted it to death with a poker - resourceful women, these West Coasters. The tale of an attack by a thylacine that was five feet in height must surely be pure fantasy.

The engraving at the beginning of this chapter was published in the *Illustrated Sydney News* on 16 April, 1867, accompanied by the following article, that shows a cornered tiger will fight. The engraving depicts the intrepid rider attacking the tiger not only with a stock-whip but also with stirrups:

"Of the Tasmanian feroe, perhaps the most savage is the native tiger. This animal, a faithful representation of which is given by our artist, is now very seldom met with in the settled districts of the island, the ravages which it commits amongst the flocks whenever it falls in such company having made it an object of the special animosity of the colonists. It bears no resemblance in form to the great feline whose name it borrows; but its skin is marked with transverse stripes, and to this particularity it, of course, owes its appellation. The gentleman whose adventure is pictured in our engraving was looking for horses on the Blackboy Plains, near Fingal, when on passing a flock of sheep, at about 300 yards distance, he observed an animal which he took to be a dog of kangaroo breed trotting through the flock. He passed on, the thought striking his mind that the supposed dog had a rather singular appearance, when, turning suddenly round, the identity of the animal with the native tiger was betrayed to him by the marks on its skin. A hunt of a most exciting kind was immediately

CHAPTER EIGHT — TIGER TALES

improvised, the tiger leading its pursuer through lagoons and timber, and maintaining tremendous speed. After a chase of an hour and a quarter the tiger began to show signs of weariness, and his enemy being enabled to come within flagellating distance, soon brought it to bay by means of a liberal application of the stock-whip. Although by nature a coward, the savage fights with an uncommon ferocity when driven to close quarters. In this instance the tiger carried on the combat with amazing fierceness; but human strength and human contrivance were against the brute, which, after a quarter of an hour's desperate resistance, fell dead under the blows of its antagonist, administered as depicted in our engraving".

Below: "The fellow had just made a kill and his face was covered in blood".
Next spread:
"Miss Murray was doing her washing in a west coast creek when she was attacked..."

161

TASMANIAN TIGER — A LESSON TO BE LEARNT

CHAPTER EIGHT — TIGER TALES

It is unlikely that, unless cornered, a Tasmanian tiger would tackle a human, although there is a record of an attack at Jericho in the Midlands, reported in the *Sydney Morning Herald* on 22 May 1872. This incident describes a man walking towards the Lake Country when he was attacked by a large tiger which came out of the scrub. He kept it off with a stick. The tiger was five feet high. This account does not really ring true, perhaps the man had too much to drink. None of the trappers interviewed ever spoke of a tiger attacking a human.

Tasmanian tigers may be inquisitive animals as several accounts tell of them following humans for some time through the bush. George Smith of Zeehan said that in 1934 two tigers hung around his camp at the Spero River and were there every night for a fortnight.

"I was in bed in daylight when a thylacine appeared..."

A remarkable story was told by Harry Walsh in 1979. In 1929, aged six, he was living with his

SMITH AND CARTLEDGE

Remind their Friends and the Travelling Public that they

HAVE OPENED A

COMMODIOUS HOTEL

AT

WHALE'S HEAD,

The Port of the Balfour Copper Field, where Visitors will find nothing but the Best. Tourists in search of fun can be accommodated with Kangaroo Hunting, Tiger Shooting, Fishing, Yachting, &c., &c. Horses provided when necessary

COMMUNICATIONS PROMPTLY ATTENDED TO.

parents in *"a poverty-stricken place called Pelverata in a large barn-like house with a very wide-opening door. I was in bed in daylight when a thylacine appeared. He wasn't shy - he just walked straight up to me and poked his nose in my face and I patted his head. He just stood there sniffing the big fire that we had. Then all hell broke loose, Mum and Dad panicked and the animal got away..."*

Tasmanian tiger hunting was promoted as a sporting activity by the hotel at Whale's Head, now Temma, West Coast. This advertisement appeared in the *N.W. Chronicle* from 1 January 1909 to 23 September 1912. It is very doubtful if any tigers were shot because the population had crashed by that time.

Tigers never offered much in the way of sport, although it was a change from the daily routine on Woolnorth for everyone to go off and "shift a tiger". The opening in 1910 of the Whale's Head Hotel at Temma was heralded by a newspaper advertisement including tiger shooting as one of the recreations offered to patrons. There is no evidence of any takers, which is not surprising, knowing the country and its weather. This is the only record of an attempt to use the animal for sporting purposes.

TASMANIAN TIGER — A LESSON TO BE LEARNT

Chapter Nine
In captivi

ty...

It could be expected that the rarity of the Tasmanian tiger and its remarkable resemblance to the wolf would have led to demand for specimens to be displayed in zoos throughout the world. A number of zoos did at one time or another have thylacine specimens, however only London Zoo maintained tigers in captivity for any length of time. Local dealers were always ready to supply animals to overseas interests, but very few were in fact exported, given the difficulties incumbent to shipping them abroad.

Firstly, a specimen had to be captured, then survive post-capture trauma. Indeed, very few specimens lived beyond this stage. The next requirement was to accustom the animal to captivity and new types of food. Only after several months would it then be ready for shipment. The advent of steamships and the opening of the Suez Canal made for quicker transportation of live animals to Europe and therefore more chance of survival. The earliest display of Tasmanian tigers in a zoo was in 1843, when the Governor of Tasmania, Sir John Eardley Wilmot, maintained three specimens in a small zoo at Government House in Franklin Square. Sir John was recalled in 1846 and there ends any account of tigers held in captivity in that collection.

In captivity, Tasmanian tigers did not provide much entertainment, apparently doing little to please visitors, looking stupid and seemingly bored with their situation. In spite of this, it is clear from advertisements published at the time that some zoos were keen to have tiger displays, and attempts to supply specimens were made (*Hobart Mercury*, 1874).

In 1994, Moeller listed 12 features of the Tasmanian tiger that he considered to be of importance with regard to the attraction, or lack thereof, for captive animals. He noted

Previous page: **A fine photo from Beaumaris Zoo, Hobart. The animal was in good condition and alert. The seventeen stripes on the back are well defined perhaps indicating that it was a young animal. The rump stripes clearly extend down on to the legs.**
Above: **Beaumaris House, Hobart, after the closure of the Zoo. The cages were to the left of the photograph.**

that they were inactive and showed no interest in either the zoo-keepers or visitors. Perhaps his most significant remark was that other mammals were more entertaining, drawing visitors away from the Tasmanian tiger, and that the rarity of the animal in no way inspired any extra attention. This was certainly the case at Beaumaris Zoo in Hobart.

A number of trappers claimed to have sold tigers to Hobart Zoo, however zoo records show no trace of this. Two animals allegedly caught on the north-west coast were dispatched to Hobart sometime between 1910 and 1912, and three animals were sent from Woolnorth by Wainwright in 1914. A further three specimens were caught at the Florentine River and sold to the zoo in 1923. Two tigers were caught in 1929, one in the Arthur River area and the other in an unknown locality. No less than five thylacines were

CHAPTER NINE — IN CAPTIVITY

also caught at Waratah in the 1930's. It was also claimed that a female and three cubs were caught in the Florentine Valley, and that these were not sold but carted around Tasmania and displayed for a charge of 6 pence per person. These were probably the animals sold to the zoo in 1923.

If all these claims are correct, then it is to be suspected that Hobart Zoo was exporting tigers on an exchange basis, despite there being no corroborating reports of acquisition from any overseas zoos. In the 1910-12 period, four tigers were sold to J. Harrison of Wynyard for £4, and sent to Melbourne. In the Queen Victoria Museum File, it is reported that they were allegedly captured by Rowe, but what actually became of them is unknown. Moeller also reported that 8 Tasmanian tigers were displayed in Adelaide Zoo, 1 in Antwerp, 4 in Berlin, 24 in Hobart, 2 in Cologne, 2 in Launceston, 22 in London, about 15 in Melbourne, 4 in New York, 2 in Paris, 1 in Sydney and 5 in Washington.

Some evidence of dealing exists in the Melbourne Zoo records held at the Records Office at Werribee, Victoria. The notes show that between 1875 and 1925, at least 17 Tasmanian tigers were obtained by either the Zoo or the Anatomy Institute at Healesville, now the Colin Mackenzie Sanctuary. In 1886, Melbourne Zoo had five Tasmanian tigers and in 1893 paid Le Souef's travelling expenses to Tasmania, perhaps to cover payment for an animal, and perhaps to obtain more specimens. Le Souef was also associated with Taronga Park Zoo, and the question may be asked whether or not the Sydney zoo was also trading in thylacines. Records also suggest there may have been regular trade in live animals coming from Woolnorth.

Another tiger in Beaumaris Zoo.

TASMANIAN TIGER — A LESSON TO BE LEARNT

Tasmanian Marsupial Wolf with three young. N. West Coast

D.C.P.
from photo.
1920.

from a photo

TASMANIAN TIGER — A LESSON TO BE LEARNT

    The extent of all this trade indicates that other zoos were dealing tigers for exchange, and in 1966 Hobart naturalist M.S.R. Sharland claimed that a polar bear, a pair of lions, an elephant and other animals were obtained by Beaumaris Zoo in this manner.

    London Zoo had the longest running display of Tasmanian tigers, and was the first to display them, from 1856 onwards. The first tigers journeyed from Tasmania to London aboard the sailing ship *Stirlingshire*, but nothing is known of how they fared on the long voyage. London Zoo had a more or less continuous thylacine display until about 1868, then there was a gap until 1884 when more tigers were on display through to 1906, then again from 1909 to 1919, with the last tigers on show from 1926 to 1931, the year in which the last tiger died in captivity overseas. The origins of this last specimen are unknown, but the Zoo purchased it from an animal dealer by the name of Chapman in Tottenham Court Road, for the sum of £150. At the time of Queen Elizabeth II's Coronation in 1953, London Zoo requested a Tasmanian tiger to become part of a collection of animals for a special display, however no permit was granted for the capture of a specimen.

Previous spread:
Painting by D.C Pearce of three young thylacine at Beaumaris Zoo.
Based on a photograph, the original of which is shown in the inset.
Below:
A pair of thylacine in the carnivore enclosure at the National Zoological park, Washington.

    The Beaumaris Zoo in Hobart had 12 tigers on display at various times after 1910 and, although all early records have been lost, it is reasonable to assume that tigers had been on display there from the zoo's opening in 1895. At least 5, possibly 10 animals were involved. Other tigers were bought by the zoo and then traded to overseas displays.

A thylacine purchased by London Zoo in 1884 lived for 8 years and 5 months, which is the longest time a thylacine survived in captivity in an overseas zoo. The animal was bought from a dealer and would have been at least 12 months old at the time of sale, which suggests a life span of $9^1/_2$ years. The Beaumaris Zoo animal purchased in 1924 did not live very long, but one of her pups survived until 1936, a longevity of some 12 years. When this animal died, its skin was offered to the Tasmanian Museum but proved to be useless because of its very bad condition.

Three thylacines obtained by the Smithsonian Institute in Washington in 1902 are of more than just passing interest, referred to as *"arriving as pouch young"*. These animals, a female with three pouch young, were sent to the zoo on 3 September 1902 by the American Consul at Newcastle in New South Wales, who paid $US 50 for them. The mystery remains as to how the animals got to Newcastle. They may have been sold by either Harrison or Beaumaris Zoo to a Sydney dealer, but there are no records of any transaction. The adult arrived in Washington in poor condition, but its health improved through to the summer of 1904 when it started to falter, refusing food and finally dying in November that year. A post-mortem revealed acute and intense inflammation of the intestine and encysted larvae of an unidentified tapeworm in all the muscle tissues.

One of the young was in a weak condition on arrival and died 9 days later. Another lived until 10 January 1905 before succumbing to haemorrhagic enteritis. The third cub survived until 13 October 1909, when it

Top:
A fine study of a male in Beaumaris Zoo.
Above:
This animal, probably a female, rests upon its haunches. This photo shows the resemblance between some of the characteristics of the Tasmanian tiger and those of the dog. Also note the large area of the hind foot which is applied to the ground.

died from fatty degeneration of the liver. The two longest surviving young are the only known examples of thylacine pouch young being successfully reared in captivity. A fascinating watercolour by Gleeson in 1902 shows the mother with three pups shown out of the pouch, although the mother's belly appears distended.

Harrison and the Beaumaris Zoo were the two major export agencies. There may have been others, as suggested by the description of the capture of a thylacine sent to Professor T.T. Flynn at the University of Tasmania, who sold the specimen to somebody in Sydney, apparently for £40. This could be the animal that turned up in New York Zoo on 26 January 1912, in which case E. Joseph would have been the likely exporter.

Many more tigers may have been exported, but there are no records of trade nor destinations. It is known that five thylacines were shipped aboard the *Ethel* in 1891 (*Mercury*, 5 December 1891). These animals were probably caught in the Derwent Valley, given that they were transferred from New Norfolk to Hobart aboard the steamer *Monarch*, but no other details have been traced.

A fine pair of thylacines in Beaumaris Zoo, Hobart. Obviously something has attracted their attention. Was it feeding time?

CHAPTER NINE — IN CAPTIVITY

Resting in the sun, Beaumaris Zoo.

Temporary local displays of tigers were organised in Tasmania, like the *"fine male specimen (...) to be seen at Warsden's Livery Stables in Hobart in the 1880's and a female with three pups (...) in display at the London Tavern in Launceston in 1864"* as reported in the *Mercury* that same year.

A Mr Bart of Tenalga bought two animals from snarers, keeping them in a shed and eventually selling them to Launceston Park in about 1912. Four other tigers were caught, the two older specimens sold for bounty and the others sold in Sydney. It would be of interest to know just how many tigers died of shock following their capture. There are reports of a tiger dying from trauma shortly after being shot in the tail, and of thylacines simply giving up when caught in a snare.

When the Hobart City Council took over Beaumaris Zoo in 1921, there was only one Tasmanian tiger on the accession list, a situation that lasted until 1923, when one animal was bought, followed in subsequent years by the purchase of an adult and her litter from Mullins, a trapper at Tyenna, and another female bought in 1925, the last recorded thylacine purchase. The Zoo lost three animals in the 1930's, the first, probably one of the Mullins' cubs, dying from *"kidney disease"*, at six years of age. The carcass of this animal was sold to the Tasmanian Museum for £5. The second specimen, a male, died on 3 July 1935, and was also one Mullins' cubs. The attending veterinary doctor diagnosed pneumonia, and the skin proved worthless due to its poor condition. The third animal,

175

A tiger photographed at Beaumaris Zoo, Hobart.

again one of Mullins' pups, was the last Tasmanian tiger to be exhibited, surviving until 7 September 1936.

The Hobart City Council records show a discrepancy between the number of animals purchased and the number of deaths. In 1921, there was only one tiger in stock when three adults and three pups were purchased. Only five deaths are recorded, and it is probable that the other two died between 1932 and 1934, a period for which the records are missing. The Zoo attempted to replenish the collection, offering £15 in 1923 at a time when £25 were required to buy an animal. The offer was raised to £30 in 1936 and a final offer of £40 was made in 1937, but no animals were obtained. The zoo declined an offer from a Mr Chaplin to catch a tiger for the Zoo on condition he was paid £4 per week for two men for a month, together with £2 for food.

The scarcity of tigers after 1921 effectively prevented the Zoo from using specimens for trade or exchange, and although there was a standing offer, no animals were forthcoming. The Government refused to issue a collection permit to the Zoo in 1937 when the £40 offer was made, and it was at this time that the first official expedition in search of thylacine was being organised.

As early as 1928, an attempt was made by some members of the Animals and Birds Protection Board to obtain protection for the species, not only because of its rarity, but also because *"prices paid by Mainland and other institutions make it impossible for Tasmanian Museums and Zoos to obtain specimens"*, as noted in the Board files.

Applications for permission to export tigers were considered by the Commonwealth Advisory Committee until 29 August 1933, when authority was transferred to the Tasmanian Government. The Board was concerned about the future of the species, and a sub-committee was established on 19 September 1933 to look into the matter. Despite this concern, a permit was granted on 21 August 1934 for the capture of a pair of thylacines for breeding in Melbourne Zoo, then another granted on 10 December 1935 to Professor Berkitt of Sydney University. Although there is no record of it, the official attitude toward exports must have altered, given that a February 1935 permit application from Belfast Zoo in Northern Ireland was refused.

There was extraordinary apathy shown by the various zoos with regard to the fate of the Tasmanian tiger. The rarity of the species and the difficulty in obtaining specimens must have been known, but no zoo expressed any concern about the fate of the Tasmanian tiger, and there are no records of any attempt to establish a natural breeding colony in Tasmania or elsewhere.

Chapter Ten
Expeditions

and searches

Out in the bush, the Tasmanian tiger will leave three clues for the searcher - footprints, dung and kill pattern. Searches have been conducted in areas where "good sightings" have been reported in recent years, in rugged country both in the search zone and on the way to the area. Helicopters have also been used occasionally to carry heavy equipment into formerly inaccessible areas.

Footprints are the most frequently sought after, but these are rarely complete or clearly impressed in the soil, and are more often than not distorted by leaves, twigs or stones. Furthermore, footprints are never a permanent feature, washed or blown away with each change in the weather, and in any case very difficult to distinguish from other animal prints, like wombat or dog spoor. Finding a perfect set of prints is therefore almost impossible, and the best impressions can only ever be classed as "probables". The days of the experienced old trappers are over, and with them have gone the skills of bushmen who could readily identify footprints. Most of the spoor sent in by people who claim to know the bush and its animals turn out to be wombat or dog prints.

The features to be examined in any spoor are the size and shape of the pad, the number of toes, the shape and size of the toe pads, the size and number of claws and whether they appear in a spoor, the spacing between the toe pads as well as the foot pads, and finally the distance between each step and the nature of the gait. Thylacine front foot spoors are different from those of the hind feet, contrary to dogs whose front and hind spoors bear resemblance. It is therefore important to have a good set of prints that enable identification of both the front and rear spoors.

Previous page: **Camp on the Studland Bay Track, where the first expedition at Woolnorth took place, in November, 1959.**

Above: **Thylacine spoor filled with plaster, Woolnorth 1960. The track is close to the station buildings.**

Prints should be examined on site, as digging up the mud or clay in which the spoors are found nearly always results in some distortion of the impression, especially when the mud dries and cracks. The distance between each subsequent print of any one foot of an adult tiger should be about 80 cm or more. This is most important in eliminating certain other species in the identification process, because two other species' spoors, the wombat and the dog, can be confused with those of tigers.

The Tasmanian devil's spoor should never be a source of confusion in this respect, since devils' pads are small and have a characteristic forefoot that leaves an almost square impression, with the toes arranged neatly in an equally spaced row along the front of the foot. Wombat spoor can be identified by the large toe pads bearing five prominent, strongly curved claws that all appear in a spoor. The wombat leaves a shuffling trail as it wanders from bush to bush, whereas the Tasmanian tiger leaves a more direct trail, heading in a definite direction.

It is difficult to differentiate dog and tiger spoor, both species travel in the same fashion and have the same gait. The plantar surface of a dog's foot is always sub-triangular with the apex pointing forward. The median pair of toes is some distance in front of the lateral toes, giving a "two up-two back" appearance. The toe pads are large and the

Possible Tasmanian tiger dung, Woolnorth. The knife is 22.5cm long. The dung contained calf hair.

median toes are quite advanced in relation to the plantar pad. The claws on all four toes of the dog are larger than the Tasmanian tiger's.

Droppings are the second source of evidence of thylacine in the field, but there again the reliability is questionable, because the tiger droppings resemble both Tasmanian devil and dog dung.

The third clue is the Tasmanian tiger's characteristic kill pattern. The throat and chest are opened, then the heart, liver, lungs and other vascular tissues are eaten together with some meat, perhaps from the inside of the ham. The nasal tissues are often eaten as well. The kill is "clean" with no unnecessary biting or savaging of the prey, however the carcass is usually consumed by devils shortly after the thylacine kill.

Many of the places where tigers were caught in the 19th century have been cleared for pasture, and on the eastern side of Tasmania much of the suitable tiger habitat is now grazing or logged forestry land. In western Tasmania, there are still large areas of inhospitable, sparsely populated and rugged country where thylacines could still survive, and indeed most searches have been conducted there. Apart from the physical difficulties of coping with the terrain and vegetation, there is ever present rain, which turns even the best trail into a horror of mud and a slippery nightmare on the steep slopes. Snow is often a hazard in the high country and can halt all progress. Heavy rain transforms rivers into raging torrents, and creeks become dangerous and impossible to cross.

The rain forests have few trails running through them. Where there is understorey, it consists of trees, cutting grass and bushes, usually very dense, making progress difficult. In parts, fallen trees form an impenetrable barrier, as their branches continue to grow skywards. Moving through this growth is probably the most daunting task of any bush work, daily progress is measured in metres rather than kilometres. There is often no understorey in the depths of the rainforest, where the ground is covered in ferns, mosses and decaying remnants of former giants. Trails are non-existent, none of the larger marsupials live there, the whole place is deep in gloom and maintaining direction is very difficult.

The rain forests are not the only difficulty for searchers to overcome. The deep gullies are by no means easy going, covered with thick vegetation and heathland plains in flatter areas. The plains consist of tussocks of a spear-like grass usually surrounded by water-filled runnels. Covering this terrain on foot is not easy, and few mammals live on these plains. The gullies are even more of an impediment, with steep rocky sides, often a deep muddy creek in the bottom and bushes that grow in thick tangles. The open forest is easier to contend with, although gullies are also present. The easiest walking is found along the west coast beaches, however the rivers cause problems, and the hinterland is rough country.

Sheep believed to have been killed by thylacine, Broadmarsh, September 1957. There is no sign of any damage to the carcass except around the head.

This is the real Tasmanian bush, a harsh environment that is hard going, even for experienced searchers carrying provisions and equipment in a heavy pack that catches on every single branch or tree, despite the increasing numbers of trails and tracks. Nowadays, most people are aware that there is an animal on the verge of extinction in Tasmania, but would probably not recognise it even if they saw one. There is confusion about the name Tasmanian tiger or devil; tiger cats are an awesome sight and could be taken for a thylacine, just as easily as a wild dog, especially in poor light. In the past, reported sightings hit the headlines before a competent interviewer had the opportunity to ask a couple of perceptive questions about the incident. Standard questioning can be carried out nowadays, however most of the eye-witnesses are now either dead, too old or unable to be located.

Most of the sightings recorded by S.J. Smith

Another sheep believed to have fallen victim to a Tasmanian tiger, Nive River-Bronte area, July 1987. The carcass was undamaged except that the top of the skull was removed with the skin turned back as far as the neck. The brain had been eaten and there was no blood on or near the carcass.

occurred in areas of low-lying ground up to 200 metres in altitude (145 out of 320 incidents) with only 43 occurring above 600 metres. This should not be interpreted as indicating an altitudinal preference by thylacines, but more likely due to the combination of human activities and road distribution, which happen to be largely concentrated in the valleys. Indeed, the species has a known distribution throughout both low and high country, although most of the carcasses presented for bounty payment were caught on the central plateau or in the foothills of the Ben Lomond massif.

Smith found that the sightings were reported in any season, by people from every walk of life. Most sightings were made by residents of the area, slightly more than a half were alone at the time of the sighting. On one occasion in Trowutta, two carloads of people spotted a thylacine at the same time. Smith's analysis showed that the tigers were only observed for a couple of seconds, that many were spotted at night and had therefore come quite close to the viewers or the road as cars went past. In any case identification was difficult and often dubious.

Since 1936, it has been lamely accepted that the Tasmanian tiger is extinct or very close to it, even in the face of persistent reported sightings, some of which stand up to considerable critical examination. This is a Tasmanian tragedy and it is disappointing that

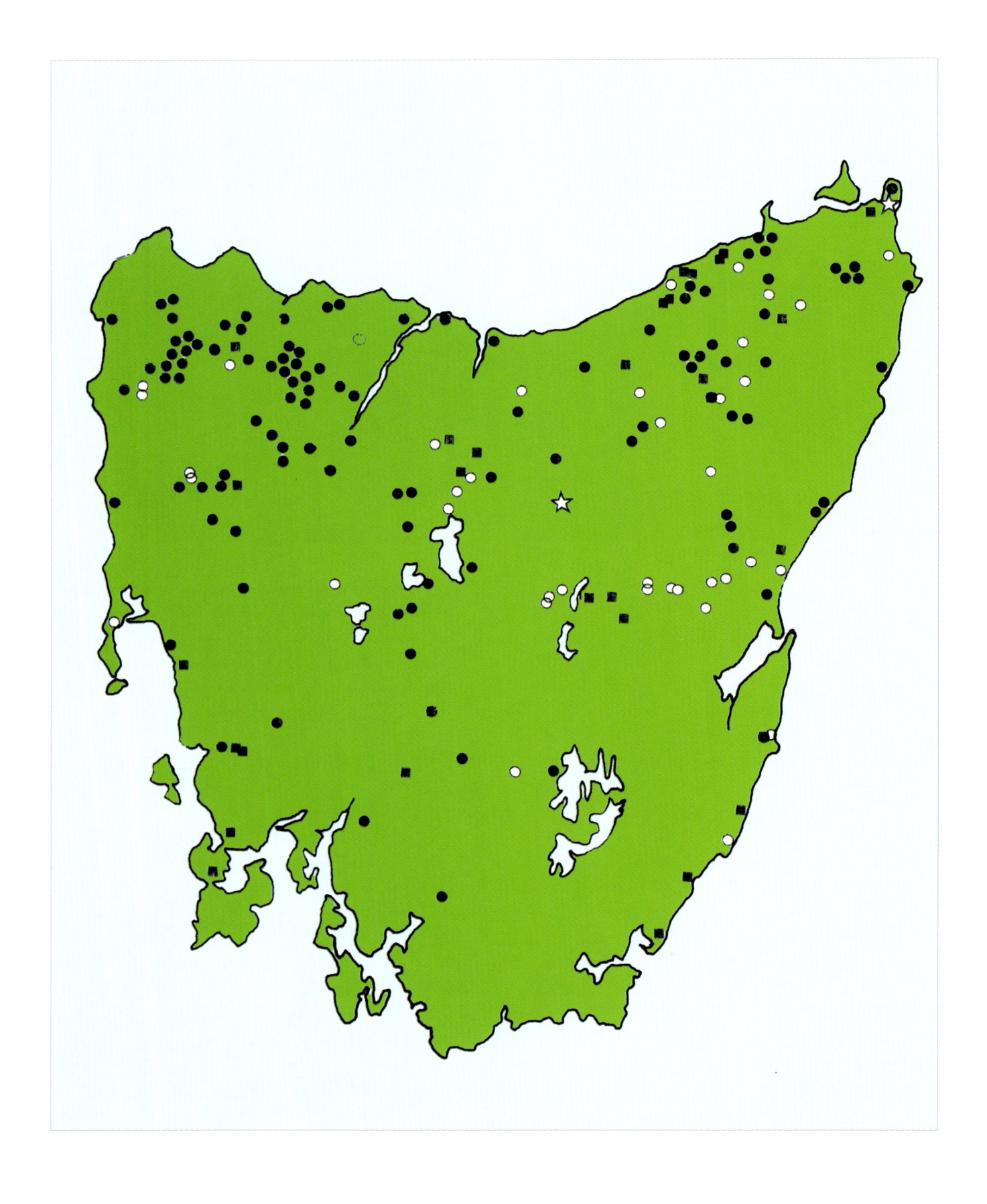

TASMANIAN TIGER — A LESSON TO BE LEARNT

no world wildlife body has sponsored a thorough search for thylacine, the rarest of all the world's mammals. It certainly seems that it is now too late.

As early as 1930, the thylacine was recognised as what is now called an endangered species, but no attempts were made to assess population or geographic spread, even though scientists in Australia and overseas must have been aware of the declining numbers.

In 1937, the Animals and Birds Protection Board, prodded by Dr J. Pearson of the Tasmanian Museum, decided to organise a major search to be conducted by the Board itself, who refused to grant permits for private individuals or organisations to seek out the Tasmanian tiger. The reason for this was to protect the animals by preventing people from capturing and exploiting them for personal profit. The only searches allowed were conducted under the Board's own rigid conditions. This attitude was not popular and although generally adhered to most searchers, unauthorised expeditions were organised. It was impossible for the Board to legislate against these activities, and in any event, none were successful in finding thylacine.

Left: Outline map of Tasmania showing locations of alleged thylacine sightings, 1936 - 1980. The stars indicate more than three incidents prior to 1970; open circles represent single incidents between 1936 and 1970; solid circles show incidents between 1970 and 1980 (Smith & Rounsevell, 1980). Squares indicate incidents between 1936 and 1985, from the author's own records. Note the concentration of sightings in areas of former abundance.

The first expedition in 1937 was lead by the Board's officer, Sergeant Summers, who set out with two others. They searched the country inland from the north-west coast, with particular attention on the Middlesex Plains where sheep killings by tigers had been prevalent in the 1830's. Finding no evidence, they moved to the coast between the Arthur and Pieman Rivers, and inland as far as the Lofty Ranges and the Donaldson River. They reported signs of thylacine presence, footprints and probable droppings. Trappers told the trio of searchers that tigers had been seen in the area, but the search remained in vain. Summers recommended that a sanctuary be set up in the Arthur and Pieman Rivers region, but no action was taken. Much of the area covered by this first expedition remains relatively undisturbed and partly inaccessible to this day.

The Fleming Searches

No doubt encouraged by Summers' efforts, the Board sent Trooper A. Fleming off with a prospector named Williams to search the West Coast Ranges in the Raglan-Cardigan River area. They set off in November 1937, and were fortunate to find tracks of *"several tigers"* on the very first day as they headed over the Raglan Range and along the Franklin River towards Frenchman's Cap. They also identified the spoor of one large and one small tiger, later finding the tracks of an adult accompanied by a young animal. Searching to the south of the Raglans, the pair found tiger tracks in 11 different locations,

Date	Description	Organisation
1937	Summer Search in N.W.	Official
1937	Fleming Search in West Coast Ranges	Official
1938	Flemming Search in Jane River area	Official
1945	Fleay Expedition to West Coast Ranges	Private
1948	Hallstrom Search in N.W.	Private
1952	Hallstrom Search in N.W.	Private
1953	Hallstrom Search in N.W.	Private
1956	Hanlon Expedition to West Coast	Official
1957	Guiler Search in Derwent Valley	Official
1958	Guiler Search at Rossarden	Official
1959	Hillary S.W.	Private
1959	Guiler Search at Trowutta	Official
1959	Guiler Search at Woolnorth	Official
1960	Guiler Investigations at Woolnorth	Official
1961	Guiler Search at Woolnorth	Official
1961-2	Guiler Search at Woolnorth	Official
1961	Guiler Search at Sandy Cape	Official
1961	Guiler Search at Lyell Highway and Bubs Hill	Official
1963-64	Guiler Expedition to West Coast	Official
1966	Guiler Whyte River Search	Official
1966	Guiler Expedition to West Coast	Official
1968	Malley Griffiths Expedition	Private
1973	Malley Various searches	Official
1974	Sayles Search in N.W.	Private
1976	Bardot Abandoned	Private
1978	Tangey Search in N.	Private
1978-80	W.W.F. Expedition, various places	Official
1982	Mooney Togari Expedition	Official
1984	Wright Expedition to N.W.	Private
1992	Australian Geographic S.W.	Private
1992	Terrey Searches, various places	Private

and thought they were made by at least four different individuals. In 1939, Fleming commented that the area had not been subjected to hunting or snaring on account of the difficult access and steepness of the gorges and gullies. The comment would still be valid today, although bushfires have taken toll on the vegetation. The area is now part of a World Heritage Area.

Not to be put off by the exhausting rigours of such activities, Fleming's enthusiasm and hope spurred the Board to organise another expedition. Again lead by Fleming, the party of 6 set out in November 1939, camping at Calder's Pass and then Thirkell's Creek, where casts were made of tiger footprints. From there the party continued on to the Jane River Goldfields, where tracks were seen in various places. Fleming recommended a sanctuary here, and today much of the zone is in the World Heritage Area. Hobart naturalist M.S.R. Sharland was with the party and wrote a detailed account of the expedition, reporting that Fleming carried a 70 lb pack for much of the time. No attempts to capture a thylacine were made on either of the Fleming expeditions.

The Fleay Expedition

The Board had granted full protection for the Tasmanian tiger in 1936, but still believed that there were sufficient numbers to allow an expedition to catch a pair for display and breeding in the Wildlife Sanctuary at Healesville (Victoria). David Fleay, then Director of the Sanctuary, set off in November 1945 with a party that included the indefatigable Fleming. The party headed to the Jane River Goldfield, since abandoned by the miners, setting snares and working extremely hard, lugging around large traps baited with bacon, live fowl and meat.

They saw no evidence of tigers and they attributed this to the fact that the area had been heavily snared from 1941 through to the time of the expedition. It was known that 2,000 snares of both the treadle and necker types had been set, the latter being particularly dangerous for the way they choked their victims. Trappers would also set strychnine baits to kill Tasmanian devils along the snare lines, and Fleay believed that these baits killed the tigers. He may not have been correct in this assumption, but was right when he considered that tigers fell victim to the snares.

Failing to find any evidence of tigers, the party moved to the Hugel lakes region of Lake St. Clair National Park, where a sighting had been reported the previous year. Nothing was however achieved here, but two sightings in November 1945 sent the expedition scurrying to the Collingwood River in January 1946. Snares were set and traps baited with delicacies such as liver, live sheep and wallabies, all to no avail.

Left: **Tasmanian tiger searches and expeditions, 1937 to date. Minor events such as brief investigations of sightings are not included.**

The expedition moved back to Jane River two months later, finding one possible footprint, hearing a peculiar cry at night that was described as similar to a creaking door. The traps were again set, and according to Fleay, a tiger was nearly caught, *"leaving a tuft of hair in the door of the trap"*. Attempts were made to lure tigers to the trap by dragging livers, hearts and various other morsels around the bush. Given the summer heat, it was highly probable that the lures were cooked - but cooked or not, they failed to attract any tigers. After a few more weeks during which a wallaby was killed right next to one of the traps, a rather disconsolate group left the traps in place, where a few local residents kept them in operation for a while longer.

Tribute is to be paid to the early expeditioners, particularly Fleming and Fleay. They endured tough and at times utterly miserable conditions, and all this without the

XSQ563 MO493 MONTREAL QUE 298/297 8 1819 COUNT PCTNS

HUBBARDNEWS BRISBANE

194 THANKS YOUR 840 COLON WE ACCEPT PROPOSITION THAT FOR
DOLLARS3000 WE ARE TO GET EXCLUSIVE GOVERNMENT TAKEN PICTURE OF
THYCALINE HUNT STOP IF NOT WOLF EVER UNCOVERED WE OWE GOVERNMENT
NOTHING STOP WILL YOU TELL THEM WE INTERESTED THEIR GETTING GOOD
STILL BLACK WHITE PICTURES AND COLOR PORTRAITS WHEN ANIMAL
CAUGHT OF WHAT THEY GO THROUGH IN ORDER GET WOLF STOP WHEN AND
TIME BRING IN ANIMAL WE MAY SEND HUTCHINS UP FOR DAY TO BRING BACK
GOOD COLOR PICTURE OF ANIMAL AND OTHER PORTRAITS HE CAN DEVISE IN
A DAYS SHOOTING TO LEAD OFF STORY STOP FOR TYPE PICTURES WE WANT
HERES GUIDE TO GIVE THEM COLON LIKE SEE TRACKING PARTY SET OFF
SEMICOLON WHEN THEY REACH PACKHORSE STAGE WE WILL WANT SEE THEM
HACKING THEIR WAY ALONG THROUGH DERBRUSH AND GENERAL WILDERNESS
SEMICOLON WE WILL WANT SEE HELICOPTER MAKING DROP IF IT DOES
SEMICOLON WE WOULD WANT ONE PICTURE OF THEM FOLLOWING
DISTINCTIVE SPOOR DROPPINGS SEMICOLON WE WOULD WANT SEE THEM SET
OUT SUCH LURES AS SHEEP AND LAMBS SEMICOLON WE WOULD WANT SEE ANY
CONSTRUCTED TRAP THEY MAKE AS THE CONSTRUCTING IT SEMICOLON WE
WOULD WANT SEE ANY ANXIOUS VIGIL THEY KEEP IN SEMIDARKNESS WHEN
THEY THINK THERES FAIR CHANCE GETTING ANIMAL COMMA THAT IS LYING
BACK IN UNDERGROWTH SEMICOLON NATURALLY ONE ENTICED INTO TRAP
WE WOULD WANT HAVE PICTURES OF THEM AS THEY LOOK AT WHAT THEY
HAVE AND TRUSS IT AND CAGE COMMA AND ALSO PICTURE OR TWO OF THEM
TRUDGING BACK WITH CAPTIVE WHEREVER AND HOWEVER HE BROUGHT BACK
STOP PLEASE STRESS ITS HARD FOR US PICK UP FROM MOVIE FILM EVEN
BLACK AND WHITE FILM SO WOULD RATHER THEY TOOK MORE STILLS STOP
CABLING MONEY STOP CAN EXPEDITION CABLE US IMMEDIATELY IF
CAPTURE MADE ?

...CATURANI .

CHAPTER TEN — EXPEDITIONS AND SEARCHES

advantages of the light-weight gear and waterproofs of today. These men had to walk everywhere with their load, whereas today, a four-wheel drive vehicle will get close to most of the places visited.

Hallstrom Expeditions 1948, 1952, and 1953

Very little is known about two of these expeditions, as they were conducted with some degree of secrecy. In Sydney in 1950, an announcement appeared in the *Herald* on 8 April, stating Sir Hallstrom of Taronga Park Zoo intended to send an expedition to catch a Tasmanian tiger for the zoo. There is no record of this expedition in the minutes of the Animals and Birds Protection Board and it is unclear where the Hallstrom expeditioners went searching, or even if they arrived at all.

Some light was thrown on the matter in 1987, when Jordan stated that he had been a professional snarer on Hallstrom's first expedition in 1948, with a party of four. Jordan did not reveal the locality of the search, but it can be inferred from elsewhere in his book that it was in the Pieman River area near an old camp where one night Jordan's brother-in-law had allegedly caught three Tasmanian tigers. The expedition may have taken place in late 1948, as Jordan complains that his permit was cancelled by the Board in February 1949, following Hallstrom's return to Sydney. Jordan goes on to make the remarkable claim that he had to release two tigers he had captured in a pit trap after Hallstrom had left.

Hallstrom announced a second search in 1952 (*Mercury*, 26 July), with the same objective of catching a Tasmanian tiger for the zoo. The permit application was supported by Turnbull, then Minister for Health. The Board was well aware of the thylacine's precarious

Left: Telegram showing the Tasmanian Government or its agent was prepared to sell the rights to any film taken on the proposed 1956-57 expedition in search of the Tasmanian tiger. It makes it clear that the publicity demands were going to control operations and that an animal was to be 'trussed up' and put in a cage, although this was never the intention of the search.
Below: Suspected tiger lair, Rossarden. December 1958. The lair was below the overhanging cliff in the middle of the picture between the tree and the cliffs in the foreground.

situation, and wished to maintain full control over the expedition's activities and captures. The *Mercury* reported that the Board did grant a permit, however there is no confirmation of this in the archives. The Board heard nothing more, save a letter that was received from Fleay requesting permission to join the search. The expedition apparently took place and in 1961 Hallstrom declared that his son had spent three weeks *"in an area where tigers were reported"*. In places where footprints had been seen, he used wallaby meat baits, some of which was taken by devils and others by an animal with different footprints. A dog that accompanied the expedition was scared stiff at night, but no tigers were caught.

The Board became more and more sensitive to people operating without their knowledge or approval. By the mid fifties, it determined to control all tiger searches and started its own investigations. But this did not stop another Hallstrom expedition in 1963 in the Arthur and Pieman Rivers region. The expedition returned convinced that Tasmanian tigers still existed[1]. This expedition was the first to be opposed by protectionists who said that if any thylacines were still around, they should be left to breed undisturbed in the wild.

The abortive 1956 Board Expedition

Encouraged by a number of reported sightings, the Board decided to launch its own expedition, and set up a committee to look after the organisation. While the expedition was in the planning and proposal stages, Al and Elma Milotte arrived from the Disney Studios in the United States, with a project to make a film about the Tasmanian tiger. They had just completed the *African Lion* for Disney, and the Board was pleased to have such an experienced team to assist in their work. The Milottes co-operated with the expedition's organising committee, but a difference arose over the copyright of the film, Milotte insisting that it would be the property of Disney, whereas the Board was equally emphatic that it should retain the copyright. Fortunately the matter was resolved when the Milottes realised the difficulties in making a film about an animal whose whereabouts, biology and very existence were unknown.

Right: **The humorists have had their fun about the searches! The caption says: "Better hide the dogs, Pa — there's another one of them city fellas comin' up here".**

The Board plodded on with its plans, but a new factor had to be taken into consideration when N. Laird of the Government Film Library joined the committee in 1956. Some secret negotiations were underway, and the Board was greatly distressed and annoyed when it heard from two sources that the Government had sold the film or photograph rights in 1957 to *Hubbard News* of Brisbane for $3,000. The telegram reads as if the expedition was being run for the benefit of *Hubbard News*. More information was subsequently revealed in a memo dated 6 March 1957 from the Secretary of Lands to his Minister, which stated that *Life* magazine wanted exclusive rights to the photographs, for

CHAPTER TEN — EXPEDITIONS AND SEARCHES

a sum of $500. Laird was consulted and said that $3,000 would be a better deal. *Life* agreed to this for the exclusive use of the photos for six months, and this arrangement was approved by Premier Reece without consulting the Board. The Board then exerted pressure on the Premier, who eventually advised that any arrangements concerning the photos had been cancelled.

These events were of prime importance in changing the Board's attitude regarding prospective thylacine searchers. The considerable financial reward to be made from a photograph alerted the Board to the possible activities of self-seekers and entrepreneurs. The value of a photo in 1958 is nothing compared to what it would be today, however the Board decided to maintain its 1937 position with regard to expeditions. In the meantime, the expedition planned in 1956 did not get off the ground, initially postponed because of the very wet winter and then finally cancelled when party leader Sergeant Hanlon fell ill. In actual fact, the Board members were so furious about the goings on with the photos that the decision was made to keep all arrangements to themselves.

Persistent reports of sightings in the Whyte River area on the Corinna Road led the Board to organise the first official search for a thylacine. Corinna was the name given to a former mining town and is an Aboriginal word for the Tasmanian tiger. Whether the name was chosen by chance or because thylacine were known to inhabit the area remains

Tiger trap in position, Rossarden. It closely resembles the trap used in 1823 at Mt Morriston.

a mystery. Sergeant Hanlon and one of the Board members, P. Wigg, spent a week in the region in April 1957, searching the river and the slopes of Mt Donaldson. Both were experienced bushmen, and Wigg had spent most of his life as a snarer, but they found nothing.

Derwent Valley Investigation

In late August - early September 1957, officers from the Agriculture Department drew attention to two sheep killed in a rather unusual fashion on the Walker property at Tanina, near Broadmarsh. It was discovered that the sheep had been killed during snowy weather, and that a neighbour had lost two sheep in similar circumstances two years previous. All the sheep had their throats cleanly torn out, the nasal bones had been eaten away and the remains of the carcasses were intact. There was no blood on or around the carcasses, which suggested that it had been lapped up by the killer. Nor was there any sign of a struggle which indicates a quick kill, and no evidence of disease was found. No wool had been ripped away, there were no signs of *"worrying"* that would indicate a dog or several dogs were responsible. Suspicion grew that the killer may have been a Tasmanian tiger, then further field investigation on the property in October 1958 revealed a lamb with the left-side ribs and liver eaten away.

W. Pearce, an old and experienced trapper who had caught tigers for bounty, said later that the method of killing was characteristic of the Tasmanian tiger, and he had no

CHAPTER TEN — EXPEDITIONS AND SEARCHES

doubt that the sheep and the lamb were victims of a thylacine. The investigation continued and it was discovered that a tiger had been sighted on a property a few kilometres away from Tanina. The owner had several visits from the animal and on one particular occasion he heard a rattling in the garbage can and plainly saw a tiger holding a loaf of bread in its large gape. The man watched the tiger for about two minutes and was sure it had stripes, but could not say how many. He described the animal as *"turning in a piece like a ship"*; he never saw the animal again. A trap was constructed and kept baited at the site until September 1958, but only devils and native cats were caught. The property was subsequently sold, the trees cleared and pastures laid.

In 1973, Dr R. Brown refused to accept that these killings were the work of a thylacine, attributing them to a dog. He stated that a large Alsatian dog had been caught in the trap at the site some four months after the last killings. The Alsatian belonged to a farmer and was kept tied up.

Taking into account all the circumstances, the methods were not typical of a dog. Dogs are known to be *"messy"* killers, and will often severely damage their prey before the actual kill. Farmers know and recognise dog kills, and Walker would have had experience of such things. The area, although close to Hobart and New Norfolk, had produced Tasmanian tigers back in the bounty days, with no fewer than 11 claims paid in the Broadmarsh Valley. It is not far across country to the Mt Field National Park-Florentine Valley district, and much of this distance could be travelled through light scrub and woodland.

The trap was large and cumbersome to put in position.

The Rossarden investigation

On 30 September 1958, Inspector MacIntyre informed the Board of the suspected presence of a Tasmanian tiger in the Rossarden region, in the foothills of the Stack's Bluff-Ben Lomond massif. A contractor by the name of Blacklow had seen footprints on a logging track outside Rossarden, and he covered the spoor over with bark. There were eight prints in all and five of them were "plastered". Only one print was particularly clear, the front paw of a Tasmanian tiger. The length between the strides was 29 inches (73.7 cm), indicating an animal three-quarters full grown. Jim Blacklow was granted permission to try and catch the animal, and a trap was sent to him by rail from Hobart. There was no point in attempting to conceal what was happening, when one of the railway porters saw the trap, he immediately realised what was going on and was reported to have commented: *"Are you bastards going to trap a tiger?"*

The trap was baited and left near a cave where the tracks indicated that a large animal had been resting. Other caves nearby revealed devils' droppings, wombat prints, and other unidentifiable spoor. For lures, Blacklow nailed sheep heads to trees and all these were eaten by some animal, including one head that had been nailed about a metre off the ground. The claw marks of some large animal were found on the tree, and it was suspected the culprits were Tasmanian devils.

The intrepid Arthur Fleming was still after the elusive animal and took part in one visit to the trap, which was left in the vicinity for several months. In spite of bullock liver being dragged around and about to attract a tiger, no further positive evidence of the species was uncovered. The area was kept under general observation until about September 1959, when the trap was finally removed. On 5 September 1959 a nearly full grown male wombat was found dead outside one of the caves with severe throat and groin injuries, but had not been eaten. Inspector MacIntyre again sighted tracks there on 1 November 1960.

There was a sequel to these events some ten years later when Noel Sutton claimed he had seen three young thylacines in the bush a few kilometres from Fingal, not far from

CHAPTER TEN — EXPEDITIONS AND SEARCHES

Rossarden. Subsequent efforts by Sutton and several field trips failed to reveal anything. Noel Sutton was anxious to find positive evidence and continued searching until ill-health forced him to retire from bush work.

The Rossarden-Fingal area was well-known for its Tasmanian tigers, where Davies had claimed annual losses of 20% of his flock. On Malahide estate, right beside where Noel Sutton was working, fifty thylacines were allegedly killed in pre-bounty times. The nature of the country in the region has not altered much and there is no reason to assume that it would no longer suit tigers, especially as there is plenty of game there for food requirements. In March 1982, a sighting was reported near Avoca, some 28 kilometres from Fingal, and less than 20 kilometres from the Rossarden area.

In 1959, it was announced that Sir Edmund Hillary intended to search for a tiger in the south-west of Tasmania (*Sydney Morning Herald*, 3 August 1959). He had, however, no such intention, and was in Tasmania for a bush walk. The Trowutta area has, in recent years, been the scene of a number of alleged sightings, and it was near here that Wilf Batty shot the last known tiger to be killed. It was therefore no surprise when a claim was made that a tiger had been spotted on the woodheap at the Trowutta Mill in August 1959. The incident was investigated but no supporting evidence could be found either at the mill or walking along the nearby tracks. The person who had seen the animal was aged about 65 and was a snarer by trade who had caught a tiger on Valentine's peak years before. There is no doubt that he would have been able to recognise the animal perched atop the woodheap.

Some time later in the same general area, a large animal was caught but escaped from a snare set by James Malley. Inspector Hanlon and Dr Guiler went to investigate with Malley. They discovered that the animal had torn down bushes in its efforts to free itself but were unable to identify the species. It was this incident that possibly started James Malley off on his quest to find a Tasmanian tiger. It is unlikely that the animal in the snare was a thylacine, because the

Left and below: G. Hanlon setting a snare. Snaring, now illegal is a dying art in Tasmania. The strong peg in the ground has a rope from the springer passed around it. A short piece of wood, the button, is notched into the peg and also into the treadle. It is held in position by tension from the springer which also raises the treadle slightly above the ground. The noose which is attached to the springer can be seen lying on the ground but is normally concealed. When an animal stands on the treadle, the button is released and the noose tightens around the animal's leg, and the tension from the springer restrains the captive.

torn bushes at the scene do not coincide with the opinion generally shared by hunters that tigers would simply give up when caught in a trap. The animal was probably too large to be a devil, and may well have been a dog, as canines are known for the strong fight they put up when caught in a snare.

Woolnorth, 1959, 1960 & 1961

The Station diaries detailed daily activity at Woolnorth, and show that in early times Tasmanian tigers were frequently seen and hunted on the property. With the co-operation of the Station Manager, P. Busby, it was decided to search the area in the hope that a thylacine could still be there. Several sightings had been reported by station employees in recent years at the Harcus and Welcome Rivers, on the Three Sticks Run and at Studland Bay. In 1959, the property was practically in the same state of development as it was in the 19th century when tigers were caught there. Some paddocks had been cleared and grassed near the homestead, but much of the original scrub remained. The old picket fences on the Three Sticks Run were found in places and later used as sites for cameras.

Right:
Studland Bay, looking south. The rocky ridge in the middle of the photograph was the site of a possible tiger lair. Many tigers were caught around these rocks in earlier times.

The expedition set out from Hobart early on the morning of 7 November 1959, establishing camp the next day on the track leading into the Studland Bay run. The very next day tiger tracks were seen at the northern end of the Bay and snares were set in likely places. The snares were set to catch the animal by a leg, as opposed to the necker snares.

One moonlit night, the party was walking along the coastal grasslands and noticed that the game was behaving peculiarly. All the animals were acting very timidly, not moving more than a couple of metres from their cover and scurrying for shelter at the slightest movement. Such behaviour was never observed on other nights nor on future expeditions. The searchers believed that it was during this trip that they got as close to a Tasmanian tiger as on any other occasion before or since. It rained through the night and about 10am the next morning a clear tiger footprint without rain specks was discovered on a muddy patch.

Encouraged by this first expedition, a new sighting on the track to Cape Grim led to the organisation of another trip into the area. The emphasis was again on finding footprints and droppings, some snares were set up using springers. For the first time automatic cameras were deployed to try and obtain a photograph of a Tasmanian tiger. The project was experimental, to find out the practical difficulties inherent to this type of exercise, rather than being the major objective of the expedition. A 16 mm Bolex movie camera was connected to a trigger set in a gap in the fence. The arrangement worked well and some good shots of possums and devils were taken. It was of interest that one wallaby had been eaten while in a snare and had its head, heart, lungs and liver eaten, although the remainder of the body was intact.

Despite a number of sightings from other parts of Tasmania, Woolnorth was still considered the most likely place to find evidence of the existence of the Tasmanian tiger. Another expedition returned to Woolnorth in February 1961. The encouraging results obtained with the cameras led to the development of a mark-II automatic set-up. Five units were constructed, using war surplus G45 8mm aircraft movie cameras with headlights for illumination. The cameras and lights were activated by a treadle measuring about 20 x 15 cm. The apparatus was not very successful as it had some unforeseen defects that could not be eliminated in the field.

The cameras operated off 24 volts and the lights off 12 volts, so the 12 volts were tapped off the 24 volt lead with a clip and resister. The resister wire got hot under operating conditions, almost causing a bush fire. Arcing contacts were also a source of problems, notwithstanding the fire hazard. The pictures were reasonable, despite the relatively poor quality of the camera lenses, and it would have been possible to identify a tiger if indeed one had been photographed. The advantage of the system was its simplicity and low cost, which was all that could be afforded at the time. The aircraft cameras were from R.A.A.F disposal stores at a cost of £5 each. One of these cameras is now on display in the Queen Victoria Museum in Launceston.

The party trudged over much of the property but saw no signs of thylacine. The fence line on which the cameras were set is of some historic interest in itself, as it was made from wooden pickets inserted through twisted wires between the posts. Snarers used to catch game by removing a picket from the fence and setting a necker snare in the gap. Perhaps this was the very same fence line along which the Tasmanian tigers were caught for bounty.

Several short trips over a few days were made in 1961-62, to check on sporadic footprint or sighting reports. One particularly good sighting was made by the Station Manager who saw the animal at the side of the road for about two seconds, before it moved off. The other sightings were equally fleeting but in each instance the viewer was emphatic that it was a Tasmanian tiger. On one trip, whilst walking across a paddock, a largish animal that looked like a tiger was seen, stripes and all. A pair of field glasses revealed a large feral cat and the locals confirmed that it was enormous and said it had even attacked their dogs.

Investigations at Sandy Cape

At Sandy Cape in August 1961, L. Thompson and B. Morrison were engaged in catching fish to sell to crayfish boats for bait. The two fishermen were camping in one of two adjoining huts, the other was used to store the nets and catch. In the middle of the night, Morrison heard a noise in the next hut, and went out with a weak torch and a piece of 4" x 2" wood in hand. He saw two eyes gleaming and hit out very hard before returning to bed. Next morning he found a half-grown male thylacine lying dead on the floor of the hut. He dragged it outside and covered it with a sheet of roofing iron that he weighed down with a length of timber, then went off with Thompson, reporting his find to the fishing boats anchored nearby. According to Morrison, he and Thompson were plied with grog and when they returned to their camp, not only was the animal missing, but a pair of paddles had also been removed. Morrison suspected that somebody from the boats had heard the tale then come ashore and taken the carcass.

Thompson, who did some geological collecting for Melbourne University, scraped up some blood and hair together with a lot of sand, and about a week later these eventually made it to the Zoology Department at the University of Tasmania. The blood was quite rotten by this time, however examination of the hairs showed that they were not from a Tasmanian devil, being longer and lighter in colour, with the cuticular scale pattern possibly that of a thylacine. Morrison turned up in Hobart shortly after and was subjected to police questioning on the matter, and his story remained unchanged in every detail. He was taken to the museum where by chance they had a live devil in a cage. The specimen was shown to Morrison, who said he knew what it was, and that it was not the same species as the animal he had killed. In fact, he was quite scathing about it all and very strongly implied that his interviewers were taking him for a fool. The press statement made by Morrison at the time was very different. He said that the animal was ferocious and escaped from a trap in which it was held by the leg. He added that he had never seen anything like it before.

Left: **A possible Tasmanian tiger lair, Studland Bay. The floor of the small cave was littered with large dung.**

Immediately after this, Inspector Hanlon, R.G Hooper and Dr Guiler set off for Sandy Cape and found the scene exactly as described by Morrison, even to the positioning of the iron and piece of timber. The Fauna Board was concerned that the carcass might be offered for sale on the open market, and every Australian fauna authority was warned of this possibility. Some information was given to police, to the effect that the carcass had been removed by a fisherman whose name was given. It was said that the fisherman had in fact offered the carcass to Sir Edward Hallstrom of the Taronga Park Trust. Sir Edward very properly refused to have anything to do with it and the carcass was then said to have been dumped at sea off the coast of New South Wales, but this remains hearsay. The story differs from the version given by Morris in 1962, who obtained his information second-hand. Morris stated that the animal had been trapped, and was so savage that it had managed to escape.

The matter became more clouded when S.J. Smith interviewed Thompson in 1980. Thompson stated that it was definitely a devil that had been killed, although back in 1961 he had said that he was unable to identify the animal. Morrison had died by drowning in 1980 and could therefore not be reinterviewed. Thompson's change of opinion is strange and puzzling, doing nothing to clarify the situation. Smith appears to support the Thompson story and points out that there are plenty of Tasmanian devils at Granville Harbour, about 50 kilometres south of Sandy Cape. Dr Guiler does not accept Thompson's 1980 version of the story, especially as he observed Morrison was a most convincing and unshakeable witness in the face of skilled cross-questioning by police.

The whole episode stresses the point that footprints, hair and sightings convince no-one in this day and age, and only a photograph or, heaven forbid, a corpse, will be accepted as proof that the thylacine is not extinct.

The Lyell Highway Investigation at Bubs Hill

In May 1961, Hank Meerding discovered a nest in a cave near the West Coast (Lyell) Highway. The nest was still warm and had therefore been most recently used. The floor of the cave was clay with sandy patches, and the nest had been hollowed out of the floor, then lined with twigs and grass. There were claw marks on the inside made by the young as they scrambled to get in and out of the nest. The nest was quite far back in the cave and was unlikely to have been made by a bird because there were no feathers lying around. There was a lot of dung lying about and this showed that two sizes of animal had been in the cave. None of the droppings were close to the nest. The clay was too hard to take an impression, but there were footprints in the sand although none sufficiently clear to identify the animal. The droppings were up to 12 centimetres long, very large to be those of a devil. The information was obtained from an investigation in July 1961, but this proved inconclusive as it is not known which species would make this sort of nest. The next day the area to the top of the Raglan Ranges and then about 18 kilometres towards Frenchman's Cap were covered on foot but there were very few tracks of any animal to be seen. Hank Meerding's son visited the scene in 1984, but did not come up with any further traces of the animal.

Below and right: Suspected Tasmanian tiger lair, Bubs Hill, West Coast. The area was littered with dung and contained a crude nest.

1963-64 Expedition

In 1962, the Animals and Birds Protection Board considered it had sufficient evidence of thylacine sightings to organise a further search. The Board approached the Tasmanian

Government to fund the project, and Premier Eric Reece, hopeful of success and the ensuing publicity, granted £2,000 to cover expedition costs and the employ of a professional snarer.

For the expedition, snaring was preferred to the cumbersome, doubtful traps, more country could be covered, more efficiently in the remote areas. Snaring was nevertheless criticised, on the grounds that the animal would be killed in the snare, however most objections were ill-founded if the springer snares were carefully set. The only risk was post-capture shock, which the old trappers had mentioned, but could be minimised if the searchers visited the snares on a daily basis.

Five members made up the expedition party under the field supervision of Inspector George Hanlon, assisted by senior wildlife officers Reuben Hooper and Ken Harmon, professional snarer Ray Martin, and Dr Eric Guiler who joined the party as often as university duties would allow. The expedition camp was set up at Green's Creek on the west coast, but after a couple of weeks without success the party moved north to Woolnorth for the summer. Following a short break, they then headed to Balfour, inland from Temma on the coast, with much the same results as before. The same variety of game was caught in some numbers, but no Tasmanian tiger, although a tiger-like footprint was seen on a track close to the camp.

The party persisted in the area until the end of May, when general exhaustion overtook everyone and finances had almost run out. Three expedition members were injured in a car accident on the road into Balfour, and this put an end to the search. The team had worked extremely hard in wretched conditions, and the sheer grind of checking the lines everyday was a major cause of exhaustion. The area covered during the expedition included the zone that Summers had recommended for reservation as a sanctuary, in which he had seen signs of tigers. The party had hoped that a major effort would have been conclusive at Woolnorth, but it was not to be, a great disappointment to all concerned.

The members of the Fauna Board and the members of all the expeditions had always had major misgivings about what should be done if ever a Tasmanian tiger were caught or photographed. The photo would have been relatively easy to cope with, concealing the area where a picture was taken was no problem. A live tiger, on the other hand, would have been an entirely different matter. News of a capture would leak out sooner or later and once the publicity hounds got the story anything could have happened. Many hours were spent around the camp-fire at night discussing the dilemma, but fortunately (or unfortunately) the problem never arose.

Right: **A "marsupial lawn" at Balfour. The grazing and browsing creatures would offer easy prey to tigers.**

An island sanctuary for Tasmanian tigers had been suggested by Professor Flynn in 1914, and in 1966 when the Fauna Board had the prospect of acquiring Maria Island partly for this purpose, it was decided to organise another attempt to catch a Tasmanian tiger, preferably two. The obvious search zone was Woolnorth, however the new management would no longer allow searches on the property. The Board was therefore obliged to seek out another suitable area and a number of places were considered but none looked promising. The Board finally lobbied the Reece Government and Maria Island was acquired in what was perhaps one of the best moves in Tasmanian Conservation history, especially as it was being sought by business interests for imminent development as a tourist resort.

The frequency of reported sightings led the Protection Board to the incorrect conclusion that thylacines had a chance of security on Woolnorth, and might even increase in numbers on the station. Optimistically, the idea was extended to other parts of Tasmania where sightings had been reported on a sufficiently frequent basis to encourage efforts to find the elusive animal. In 1966 a new expedition received support from the World Wildlife Fund, in the form of a £1,000 grant. It was decided to search Granville Harbour on the west coast where George Smith, a man with years of bush experience, had seen footprints that looked like those of a tiger on his farm.

The expedition established a base at Granville Harbour where there were good living quarters and vehicles for transporting equipment. The cameras and snares could be serviced with greater ease than anywhere else. Hindsight would indicate that this expedition followed too soon after the previous effort, and the party members still had previous disappointment in mind, and enthusiasm had not been rekindled. There were also general misgivings about snaring for Tasmanian tigers, and the fear that one would die if caught.

In June 1966, there was talk of a possible tiger lair in an abandoned boiler at an old gold mine on the Whyte River. It was discovered by Reuben Charles who took part in the ensuing investigation with Rex Davis and Dr Guiler. The area is covered in fairly thick forest that had been logged in the past, and later cleared for mining. Old mine shafts, exploratory holes and the like presented potential hazards, however the mine access road was fit for getting to the site on foot. A trap was built with a compartment in the middle and cages on either side with drop doors operated by a treadle arrangement. Food, water and shelter were provided in the middle section for live bait that could not be killed by an animal in the trap compartments. The lures were live wallaby or hens, the latter proving productive in providing the occasional fresh egg.

A camera was also rigged up using an updated version of the mark-II gear and the G24 cameras replaced with an 8mm Halima. This mark-III version took some

This hut at Balfour served as a base camp.

photographs, mostly of pademelons and bandicoots, however one sequence of photos showed a blurred animal moving quickly in the lower corner of the frame, but the species could not be identified. The area was unsuccessfully searched for footprints, apart from those of devils, wombats, wallabies, pademelons and mill dogs. The usual oddities were caught in the trap, but overall the number of animals caught was low, and no tiger cats were captured. On several occasions the trap had been sprung by some animal large enough to activate the treadle near the bait while having enough of its rump under the door to stop it from closing, thus allowing the animal to reverse out. Some hair collected was possibly of tiger origin, but only dog spoor were found around the cage. Trapping continued for about a year, largely due to the enthusiasm of Rex Davis, but these efforts proved unsuccessful. Some 23 years later, a report of a sighting and possible footprints was made at Mt Donaldson, only a few kilometres from the Whyte River.

In March 1968, Jeremy Griffiths of New South Wales and James Malley of Trowutta set off on a search that to a certain extent was sparked by Malley's previously described discovery of something that had threshed about the bush while caught in a snare. The expedition was the biggest effort made thus far to find a thylacine, and certainly deserved more success. They walked over a great deal of Tasmania, particularly in the north-east and west coast districts, even trekking from Macquarie Harbour to Port

CHAPTER TEN — EXPEDITIONS AND SEARCHES

Davey. Malley built some automatic recorders that he installed on likely tracks. These were designed to record any passing animal larger than a devil, and the information was passed back to the base by radio. The idea was good but did not work too well due to "bugs" in the system and a lot of animal traffic.

Griffiths and Malley were joined by Tasmanian bushwalker Dr. R. Brown, and together they established the Thylacine Expedition Research Team. In due course, money was subscribed and 25 automatic camera units were spread widely over likely localities. In addition, a Tiger Centre was set up for the public to give and seek information, and this indeed uncovered some useful contributions, arousing considerable public interest. The two leaders of the expedition accomplished a tremendous amount of work in their 4-year quest, and if effort had been the key to success, they would have got results. However, like all previous searchers, they got neither photographs nor live specimens, but did manage to collect a lot of information as they went along, with enough sighting reports to encourage them in the undertaking. The team was finally disbanded in September 1972.

The expedition reports illustrate the difficulties of interpreting field data on an animal about which so little was known. Expedition members did not agree with some of the views of other searchers, in the same manner as others did not agree with them. The disagreement even extended to within the team, Malley and Griffiths believing the tiger to be extinct. They concluded that a factor in

Track to Top Farm, Granville, 1966. Coastal tracks such as this were used by a wide variety of animals.

thylacine extinction was the trappers' habit of poisoning snare lines to kill devils and these baits being eaten by tigers. However, from what is known of thylacines in the wild, this is unlikely as the tigers are not scavengers.

Throughout Brown's report, there is strong criticism of the lack of official support for the expedition. It was the Government's view that any searches should be carried out by official agencies, that private expeditions were not to be encouraged, reflecting the 1937 decision. All wildlife was, and still is, the property of the Government, and any rewards should be to the people of Tasmania through the Government.

In 1973 the National Parks Service employed Malley to check out sighting records from around the State. He investigated nine events, seven that he believed were "good" sightings, five of which were in the north-west. It seems a little superfluous to check all sighting records, in so far as the questioning of viewers was completely uncoordinated. The following year Sayles and Tangeny conducted essentially solo searches, again unsuccessful. Sayles was assisted by the ever-enthusiastic James Malley, and used a typically American invention called the "varmint caller" that squawked like a wounded animal. Sayles spent nights perched up trees puffing into his instrument, and on one occasion actually succeeded in attracting an animal that he could not identify, save that he said it was much bigger than a devil. In 1980, Sayles expressed the awful hanging doubt as to what he had seen.

CHAPTER TEN — EXPEDITIONS AND SEARCHES

The use of the "varmint caller" is interesting, but whether it would have attracted a Tasmanian tiger is another matter. Most marsupials are silent when hurt, and the thylacine is not known for a sudden cry of pain. The approach was nonetheless original, worth trying, and could have paid off.

1976 saw actress Brigitte Bardot propose to conduct a search for the Tasmanian tiger, but the project was cancelled before it even started, largely because it would have cost too much and required staff to run. Two years later, the World Wildlife Fund (Australia) was launched, and an important part of its initial programme included assistance to search for the Tasmanian tiger. A grant of $55,000 was made to the National Parks and Wildlife Service for a two-pronged approach to be conducted by the Service and Dr Guiler. The Service appointed Steve Smith to investigate recent sighting reports and conduct field operations using purpose-built automatic cameras. He interviewed a few of the remaining people who knew of the thylacine, and collected some historical information and sighting reports. His results were encouraging but still no thylacine or even a photograph were obtained, and in the absence of any conclusive evidence, he was forced to concede that the species was extinct.

Smith details his search in his large, comprehensive 1980 report. He analysed all the sightings between 1934 and 1980, and found that the number of sightings per decade rose from 20 in 1934-40 to 99 in the 1960-70 period and 125 from 1970 through 1980. One wonders whether Jordan had a flash of vision in 1987 when he predicted that the number of sightings *"would increase to mammoth proportions"* as extinction closed in on the species. The publicity surrounding the Griffiths-Malley Expedition probably flushed a lot of alleged sightings from out of the woods.

Left:
A small wallaby (pademelon) believed to have been killed by a Tasmanian tiger at Top Farm (West Coast) in 1976. The throat, neck and upper chest have been ripped out but the remainder of the body is untouched. There has been no chewing or biting on any other part of the carcass.

The sightings were concentrated in the north-east and north-west with isolated incidents elsewhere and only a few reports from the south-west where the habitat is poor and visitors are rare. Only 107 of a total 301 sightings were placed in the "good" category. 94 sightings occurred below 100 metres altitude, and but 70 were of more than one animal. Most of the sightings were reported at dusk or after dark, and more than half were witnessed by a single person for under 30 seconds. Every sighting cannot be taken as hard evidence, but in view of Smith's findings and accepting that each sighting is valid, it can be confirmed that Tasmanian tigers do not favour mountainous terrain, are nocturnal and tend to be solitary, although groups of thylacines have been seen at various times.

Smith built and operated six automatic cameras placed in different locations and different habitats. He obtained photographs of most of the Tasmanian mammals, but no Tasmanian tigers. The search was based on two years of field work using the automatic

Famous French actress Brigitte Bardot, who, since distancing herself from the movies, devotes much time and effort to the protection of animals, with a fighting spirit that earns her plenty of animosity in her own country. She also dreamt of finding traces of the thylacine.

cameras, as little was to be gained by rushing frantically about Tasmania to investigate every new reported sighting. This tactic had been adopted by Fleay in the forties, during Government searches in the late fifties and 1960's and by the Griffiths-Malley team, but all to no avail. It is likely that no new information will be gleaned now, most of the recent interviews are largely repetitive, although some historical snippets may still turn up.

Cameras placed in likely locations over long periods of time are likely to bring more chance of success, and with this in mind a considerable portion of the allocated $25,000 was spent on the construction of fifteen camera units using a prototype that had been designed for another project. Finding a suitable area in which to place the cameras was no easy exercise. Although the units required little servicing and could be left in the bush unattended for a fortnight, the area had to be somewhat isolated to ensure freedom from interference and vandalism. Good recent sighting records were another prerequisite, if possible in areas historically noted for thylacine presence.

The apparatus consisted of a pulsed infra-red beam that activated a movie camera and lights when interrupted. Each unit had a transmitter, sensor, Super-8 camera, lights

CHAPTER TEN — EXPEDITIONS AND SEARCHES

and control box which doubled as a carry case for some of the equipment. The electronics were stored in a weatherproof box in the control unit. Fifteen units, each costing about $550, were built, but in practice only twelve of these could be maintained at any one time. In the field the most severe problem was caused by cattle kicking over the batteries, chewing the leads, knocking over the camera mounting and trampling on sensors and transmitters. Despite all this, the electronics did not suffer any damage.

Tasmanian devils also chewed the leads, particularly from the sensors. On one occasion a spider spun its web and nest inside the sensor tube, and there was also some trouble caused by ants building nests inside the camera box and body. The units were exposed to quite a variety of weather conditions, including temperatures ranging from –5°C to +40°C. The cameras were also operated in regions where rainfall is high (150 cm per annum) and were twice buried under snow. Moisture in the wooden boxes caused corrosion at the terminals in the box, particularly around the photoelectric cell, but this was corrected by isolating the terminals from the box. Excessive humidity also caused corrosion of the printed circuit, and two units were written off after animals had overturned them and water found its way into the control panel.

Automatic recording camera in position on a game trail, Sawback Ranges, 1978. The control box is at the left with the infra-red transmitter tube beyond it. The sensor is hidden in the ferns on the right. The camera and lights with their battery were at the position from where the photo is taken.

The units could be left undisturbed on site for about ten days, after which routine maintenance was necessary, which included checking sensor alignment, the amount of film and the operational efficiency of the equipment. The batteries had to be checked every 20 days, in particular the 12 volt accumulator. The disadvantage of the system was in the weight of the units, lights, batteries and stand that totalled about 40 kg. A modified back-pack enabled two units to be carried, and a wheelbarrow was used to transport four units on suitable trails. Needless to say that the units could only be transported over relatively short distances, and in practice could not be moved far from tracks.

Some aging of the electronic units was experienced, particularly in the integrated circuits in the transmitters, which would stop working without warning. After three years of wear and tear in the bush, ten units were still functional.

Four units were taken to Adamsfield for field trials over four weeks. Minor electrical faults were corrected and the programme was immediately activated, with the decision made to concentrate installations in one area rather than scatter them widely. Sites where thylacine sightings had been reported in recent years were chosen.

Each site offered a different habitat type. The highest rainfall was in Adamsfield (180 cm per annum) where the vegetation is wet sclerophyll forest with open sedgelands on flat marshy areas. The cameras were set on trails leading from the forest to the plains, and later seven units were set along a track leading to Adamsfield at higher altitude where some snow fell. The west coast has more varied vegetation and is drier with only about 140 cm of rain a year. Much of the area was cleared pasture land with heavily wooded gullies surrounded by rain forest dominated by Eucalyptus or Nothofagus. Here the units were located on trails leading to paddocks.

The Mainwaring River site was located on the coastal dunes and grassy areas in the coastal scrub. There was no forest in the area, the only trees being some stunted Eucalyptus on the coast. Extensive sedgeland plains extend inland behind the narrow coastal belt. The rainfall is about 140 cm per annum but there is considerable salt spray from the very rough seas that roll in to the coast. The cameras were set on trails leading to the grassy feeding areas used by herbivores, the area was only accessible by helicopter.

The north-west zone was located inland in a region of mixed regrowth over mature myrtle forest with about the same rainfall as Mainwaring River. The area had been logged, and the camera units were dispersed along the logging tracks. Lastly, the Lyell

CHAPTER TEN — EXPEDITIONS AND SEARCHES

Highway site was beside a road with myrtle forests on both sides and 200 cm rainfall per year. There were no obvious trails running through the forest and consequently only one unit was used, at the spot of a recent sighting.

Togari 1982

In Togari in the north-west, Ranger H. Naarding reported seeing a tiger in March 1982. Naarding had previously worked with wildlife in Africa before moving to Tasmania, and his was the only sighting by an experienced wildlife worker. He was sleeping in his vehicle which was parked at a road junction south of Togari, his account reads:

"It was raining heavily. At 2.00 am I awoke and out of habit, scanned the surrounds with a spotlight. As I swept the beam around, it came to rest on a large thylacine, standing side on some six to seven metres distant. My camera bag was out of immediate reach so I decided to examine the animal carefully before risking movement. It was an adult male in excellent condition with 12 black stripes on a sandy coat. Eye reflection was pale yellow. It moved only once, opening its jaw and showing its teeth. After several minutes of observation I attempted to reach my camera bag but in doing so I disturbed the animal and it moved away into the undergrowth. Leaving the vehicle and moving to where the animal disappeared I noted a strong scent. Despite an intensive search no further trace of the animal could be found".

Left: **Setting cameras at the Mainwaring River on the west coast. The scrub offered many trails for positioning cameras.**

This sighting was kept confidential for two years and an uninterrupted investigation was carried out. The automatic cameras were set up on likely tracks and Research Officer Mooney made a major search of some 250 square kilometres of the surrounding forest, heath sedgeland and pasture. There had been reports of thylacine sightings in the area over the previous two years, and Mooney spent a lot of time preparing sandy or muddy patches on the trails in the hope of collecting footprints. A total 256 artificial patches and 89 existing ones were regularly examined, three potential dens were regularly monitored and 50 others checked out. Wallabies killed on the road were left beside the sandy patches, plaster casts were made of likely footprints and faeces were collected for chemical analysis.

No corroborating evidence was found, no footprints, no droppings, no photographs. A psychological reason may explain the phenomenon. Naarding may have had a "waking dream", which left him convinced that he had seen a thylacine in the pouring rain. In general, animals do not like heavy rain and tend to avoid it by keeping to cover, a fact that lends some strength to the notion of a dream.

This was the most thorough and wide-ranging search ever carried out but it yielded only the meagre result of several "possible" tracks. The area was used by

Tasmanian tigers in earlier times although recent forestry operations may have frightened them away, despite sufficient numbers of prey species for an adequate food supply. Mooney points out that the presence of this tiger was reported 45 years after the last specimen died in captivity. He also emphasises the difficulties in preparing a proper management plan in the absence of any biological knowledge of the species. The cameras were then handed over to the National Parks Service who subsequently placed them in a number of places where sightings had been reported in the past.

With the exception of the forester kangaroo, every other species of large marsupial living in Tasmania was photographed. No films showed a thylacine and the expedition to date must be considered a failure. A great deal of information has nonetheless been gathered about other species and the utilisation of trails. It was clear at all the sites that the trails were used by different species on the same night, and the presence of carnivores such as the Tasmanian devil and the *Felis cattus* cat apparently did not deter other species from following the trails.

There were a surprising number of animals active in daylight, when about one-eighth of all incidents were recorded, some occurring at midday as can be seen from the shadows on the pictures. The species most commonly found active in daylight were the wallaby, the Tasmanian devil and the feral cat. Neither wallabies nor devils were seen in daylight in the Granville locality and it is probable that the activity is confined to heavily vegetated areas. On several occasions an animal using a trail could be identified by individual characteristics and noticed passing along the trail in the late evening or early part of the night, returning later or next morning, presumably going from its resting place to its feeding site.

Right: The site of the Naarding sighting at Trowutta. His vehicle was parked on the Triangular part of the road at the junction of the forestry tracks. The trees are nearly all regrowth eucalypts.

In December 1983, to commemorate Condor's win in the Sydney-to-Hobart Yacht Race, American Ted Turner announced a $100,000 reward for any positive and confirmed evidence of the existence of the Tasmanian tiger. Such rewards, however well meant, encouraged searches and expeditions of various degrees of competence and would certainly have disturbed remaining specimens. At best, an expedition could turn up some evidence whereas at worst the result would be a dead animal, the very last thing anybody, particularly the sponsors, would wish to see. In any case, these private expeditions were an administrative nightmare as there was no assurance that security for the tiger and its habitat would be enforced before the news was released to the media. The spate of publicity that followed the Turner offer prompted the National Parks and Wildlife Service to issue a warning that catching, trapping or shooting a thylacine carried a maximum penalty of $5,000 and/or six months in prison.

In January 1984 it was announced that P. Wright intended to launch a search in the Western Tiers near Mole Creek, where Wright owned a Wildlife Park. A company

CHAPTER TEN — EXPEDITIONS AND SEARCHES

was formed to finance the affair and a base camp was built near Lake Adelaide, a region where there had been both historic and recent alleged sightings. Wright claimed to have reports of 50 sightings in 1982-84, but only considered eight or nine as "good". The expedition was most elaborate. The materials for the base were flown in by helicopter, and, apart from providing quarters for a team of three, the camp acted as a receiving station for information beamed in by radio from the automatic cameras. Parameters such as time, date, temperature, noises, wind and movement were also recorded. The electronics cost $50,000 but were not entirely successful because the gear was too sophisticated and had not been adequately tested under field conditions. Consequently it was full of "bugs", the batteries required daily servicing to maintain the cameras in working order, and continuous communication with the base was not possible.

In August of the same year it was announced that operations would be scaled down as the $250,000 budget had been exceeded. The helicopter was removed and the ten employed men were dismissed in the bush. The cameras were left in position for some time, serviced by Wright on horseback, until the whole project was wound up later in the year. It is claimed that the search failed because the lack of snow failed to drive tigers down to lower ground where the cameras were positioned. No detailed report was ever published. At about this time F. Flammea made a proposal for a well-financed and electronically advanced expedition over several months, but the project never got beyond preliminary discussions.

TASMANIAN TIGER — A LESSON TO BE LEARNT

Department of Lands Parks and Wildlife

44 Macquarie Street or GPO Box 44A Hobart Tasmania 7001 Fax (002) 238765

Réf. N°.

Réf. N°.

THYLACINE (TASMANIAN TIGER) SIGHTING REPORT FORM

Every observation, no matter how trivial it may seem, might prove to be important in the search for the thylacine. All information will be received strict confidence. Thank you for your assistance.

OFFICE USE
G
F
P

NAME .... PHONE N°......................................
ADDRESS...
(DIRECTIONS TO HOUSE) ....
OCCUPATION .... ACTIVITY AT TIME OF SIGHTING (e.g. shooting) ....
DATE REPORTED .... /..../19...... RECEIVED BY... INTERVIEWED
....

SIGHTING DETAILS

LOCATION : Exact location ; nearest reference point (map co-ordinates ?)..............................
(Altitude........................ m) ....
DATE ..../..../19...... TIME .... HOW MANY MADE THE SIGHTINGS ....
TRANSPORT (car, foot, etc.) ....WHAT WAS SEEN (animal, tracks etc.) ....
....OBSERVED FOR (time) ....
IF ANIMAL SEEN, DESCRIBE ITS MOVEMENT (e.g. moving left, trotting etc.) ....
....
....
WHERE SEEN (road or track/beach/paddocks, bush, etc.) ....
....
SURROUNDING VEGETATION (rainforest/gum trees/low scrub/buttongrass, etc.) ....
....
.... DISTANCE (feet, metres, chains etc.) ....
LIGHT SOURCE (headlights, sun, moon, etc.) .... WEATHER ....
DESCRITPION OF ANIMAL (head, ears, body, tail, colour, markings, length nose to tail, height) ....
....
....
....
COMPARISON WITH OTHER BUSH ANIMALS (wild dog, wild cat, devil, wombat, tiger cat, native cat) Please comment briefly ....
....
....
ANY OTHER COMMENTS ....
....
HAVE YOU SEEN A TIGER REPORT ....WHEN AND WHERE (alive, photo, etc.) ....
....

CHAPTER TEN — EXPEDITIONS AND SEARCHES

In 1992 a young couple spent a year on a survival exercise at the Wanderer River in the south-west. They were not searching for Tasmanian tigers but no doubt they were alerted to the possibility of seeing one, however none were spotted. Over the last 20 years, several alleged sightings were made in this area and the absence of animals may be significant.

In north-east and north-west Tasmania, R. Terrey has been searching for thylacine for many years, even going as far as digging pit traps in an endeavour to capture a live specimen. Terrey has interviewed many people who claim to have seen tigers, but so far his efforts have been unrewarded.

The question remains as to what, if anything, has been achieved by these searches and expeditions. Some were of short duration, lasting only a few days, others extended for weeks, even years. All failed to produce final irrefutable evidence of the existence of the Tasmanian tiger. It is abundantly clear that there is little point in plunging head first into the investigation of new sighting reports, in which identification of the animal may be incorrect and thus time and money wasted. Even when a Tasmanian tiger has been sighted, no evidence has been found to prove that the animal remained in the area. Mooney's efforts at Togari after the Naarding sighting clearly point to the likelihood that the animal was only passing through. In 1987, Jordan may have been right in stating that a thylacine would not return for months, but perhaps the species would move on after an encounter with a human being.

This form has been used for report of thylacine since the mid 1970's. It is still used with minor revisions. Such forms are a convenient base for less experienced personel to interview. Usually reporters are invited to sketch the animal on the reverse side.

On the other hand, historical evidence from Woolnorth suggests that Tasmanian tigers will stay in an area until they are caught. However, long term searches using photographic methods have proven equally fruitless, suggesting that an animal never returned to the spot where it had been seen. It would be impossible to cover all the trails in the bush and perhaps the tiger was moving around but never tripped a camera mechanism. The conclusion that the thylacine is an elusive animal is disconcertingly obvious, especially in view of the fact that not a single specimen was photographed in all the years that automatic cameras were in position. It may well be that the species had learnt to avoid human contact, but the sad fact is that so little is known about thylacine habits that one cannot even be photographed, let alone caught.

1- Everyman, 17 July 1963; Hallstrom letter to Laird, 1963 in *Laird file*, State Archives.

TASMANIAN TIGER — A LESSON TO BE LEARNT

Chapt

The im

Chapter Eleven
possible dream

The first part of this final chapter is in no way intended to be a biology lesson, but merely to provide certain basic notions, so that readers will be in a better position to comprehend the rest of the discussion, which aims to answer the fundamental question that so many people quite rightly ask: Given current advancements in molecular biology and their foreseeable perspectives, can one seriously envisage the future recreation of an extinct creature from among the many species struck from the list of those that originally inhabited our planet?... a consequence of the short-sighted policy and unequalled selfishness of the most elaborate, and without doubt most cruel, of all predators to have ever rampaged here on earth: man.

Thylacine belongs in the category of higher organisms made up of a collection of eucaryote cells, which are cells whose chromosomes do not directly bathe in the cytoplasm - such is the case for bacteria - but are enclosed in a nucleus separated from the cytoplasm by a porous membrane that allows exchanges.

All the cells of a given living organism come from a mother cell. This original cell, or zygote, is said to be diploid, as it contains two pairs of each chromosome (23 in man and 14 in thylacine). This original root, or zygote, is the result of the fusion, that occurs during fertilisation, of two haploid cells, or gametes, which only contain a single set of chromosomes. One of these gametes is the sperm from the male, the other the egg contributed by the female.

Once this fusion has occurred, mitosis, or cell division, begins. This represents the key to the development of the embryo. All the newly formed cells will in turn undergo mitosis, until a given being is formed and endowed with its various vital functions.

Within a higher organism such as the one that we are interested in, the number of cells is in the region of one million billion, which in itself is mind-boggling. With all these cells coming from the original zygote, they share the same constitution and are only progressively differentiated during the development of the foetus.

The chromosomes present in the cell's nucleus contain long macromolecules of DNA (deoxyribonucleic acid). The latter comprises a skeleton to which four nitrogenous bases are attached. The fundamental discovery that has enabled molecular biology to make astonishing advances in the last fifty years is that genetic information is exclusively dependent on the order in which these bases are arranged.

The bases are spread in linear fashion along the DNA macromolecule and are grouped in genes; the sequence is the order in which the DNA bases follow each other. Each gene comprises its own specific number and arrangement of bases. This constitutes the very information with which the cell it is part of, can accomplish a very precise chemical reaction by way of a protein with a particular function.

Previous page: **It is from ADN taken from this type of pouch specimen, preserved in alcohol, that amateurs of science fiction dream of one day mastering the process of recreating an extinct animal.**

Among these proteins are enzymes which play the role of catalysts. Without proteins, the aforementioned chemical reactions could only occur at very high temperatures (several thousand degrees) or at a rate so slow as to be incompatible with the requirements of life.

A law of correspondence, known as the "genetic code", links a given gene to a given protein. This law is common to all living things, including plants. All the cells of a given individual contain the same genes. However, if the mere fertilised egg, from which that fantastic adventure called the genesis of a new higher being commences, leads to the production of an extremely complex organism with a multitude of organs, this is because certain genes are activated, at a given time, whilst others remain inactive. From there comes the specific nature of each cell (liver, heart, brain, bone, etc.). The mechanism that constitutes this differential activation has yet to be entirely elucidated.

The characteristics of every living thing are therefore found in its genes, with all the genes constituting the genome of the species in question. To date, it has not been possible to totally decipher man's genome, although the international scientific community agrees in thinking that it is only a question of time and means, and that the future is not too distant when every higher organism will, one after the other, be able to have a genetic identification card, every gene having been located in sequence and the role of each one of them understood. This has indeed been recently achieved for an archaebacteria, and is about to be for a worm on which many scientists are working, as well as for brewer's yeast.

Henceforth, dreams become possible and some people have indeed started dreaming ever since a new science, called genetic engineering, has enabled manipulations to be carried out on genes, and the possibility to intervene in the very destiny of an individual at the embryonic stage. Accordingly, over the last twenty or more years, biologists have had at their disposal a series of biochemical instruments with which to carry out surgical interventions on DNA.

Restriction enzymes play the role of real knives, making it possible to cut DNA in specific places, and two strands can be joined in a lasting manner. In this way it is possible to produce composite DNA molecules using DNA of two different origins.

This possibility has led to the gene grafting technique, using a viral vector to infect a target-cell, which has opened up new horizons for medicine to correct genetic disorders. It also means that, in the relatively short term, the partial - the word punctual would without doubt be more appropriate - modification of a higher organism can be envisaged, by modifying the egg in the immediate post-fertilisation stage.

Such a possibility immediately gave rise to exclamations of a miracle, and some have even gone a step further, leading the general public to believe that scientists were on the verge of being able to give life back to extinct species, so long as one possessed their genetic I.D. card. This belief was recently reinforced by news that biologists had managed to extract DNA from Egyptian mummies a couple of thousand years old!

Before proceeding any further, however, it must be noted that this DNA is very rarely intact, in other words a number of sequences are missing; not to mention the frequent contamination (astronomical quantities of DNA float in the air of laboratories) giving rise to absurd results - a pig bone for example, several centuries old, able to supply human DNA, and vice-versa. In general terms, biologists consider most uncertain the extraction of DNA from remains - bone, skin, dried tissue - of higher organisms more than a century old.

However, in the case of the thylacine, that we are specifically interested in, the question arises in more encouraging terms since, without speaking of "fresh" remains, several embryos (pouch specimens) conserved in alcohol, do exist. In itself, alcohol is of a nature to enable good conservation of DNA. However, the DNA extracted from an embryo conserved in such a way could well find itself very closely associated to denatured proteins, which would make development difficult.

But let us come back to the subject of manipulations enabling scientists to play apprentice-sorcerers creating pipe dreams. The strategy for such manipulations, already carried out on drosophilus flies, is as follows: a specific gene, responsible for one of the characteristics of the species - the colour of the fur for example - would be taken from the remains of an extinct higher species. Purified by cloning, this gene would be introduced

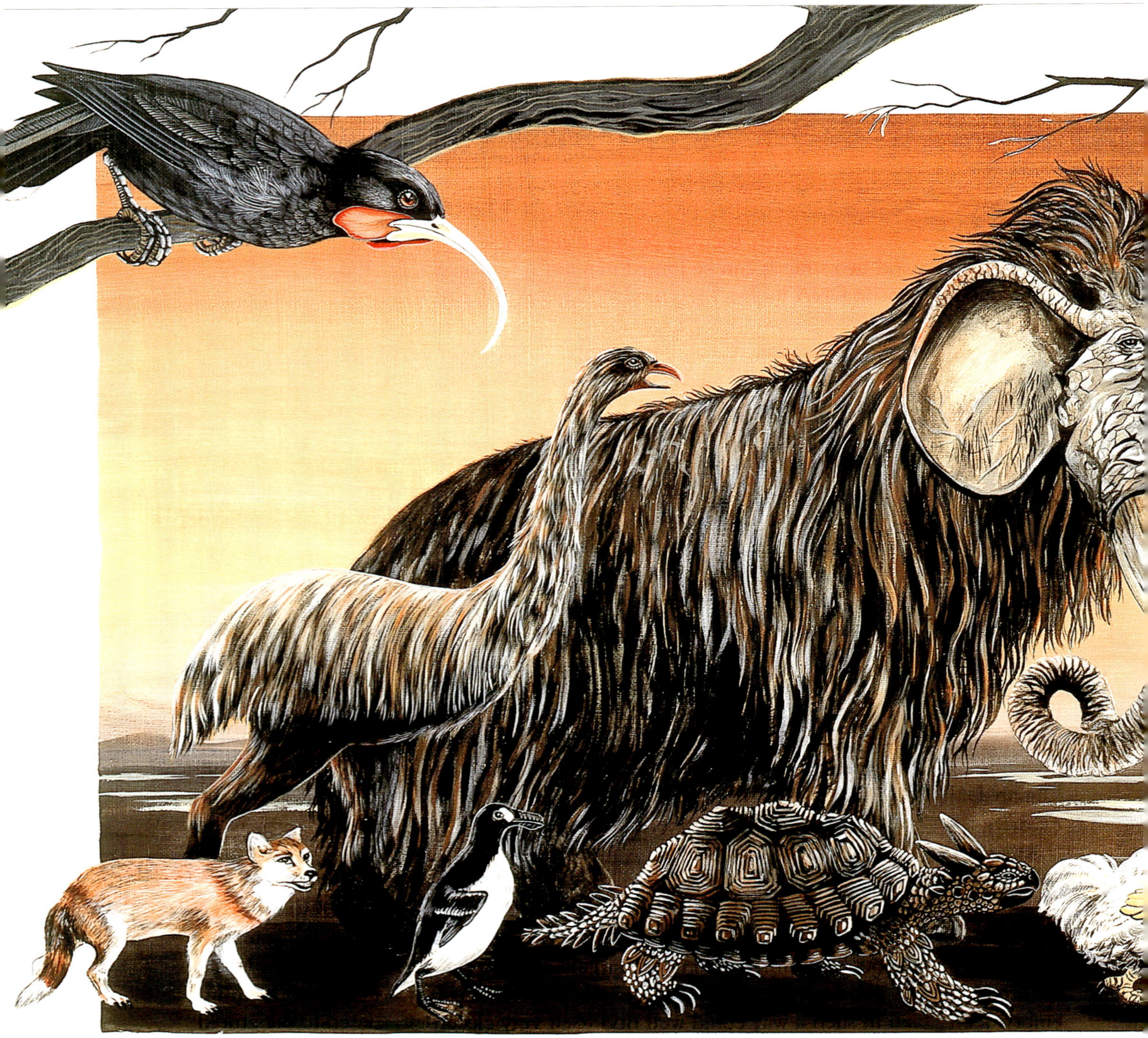

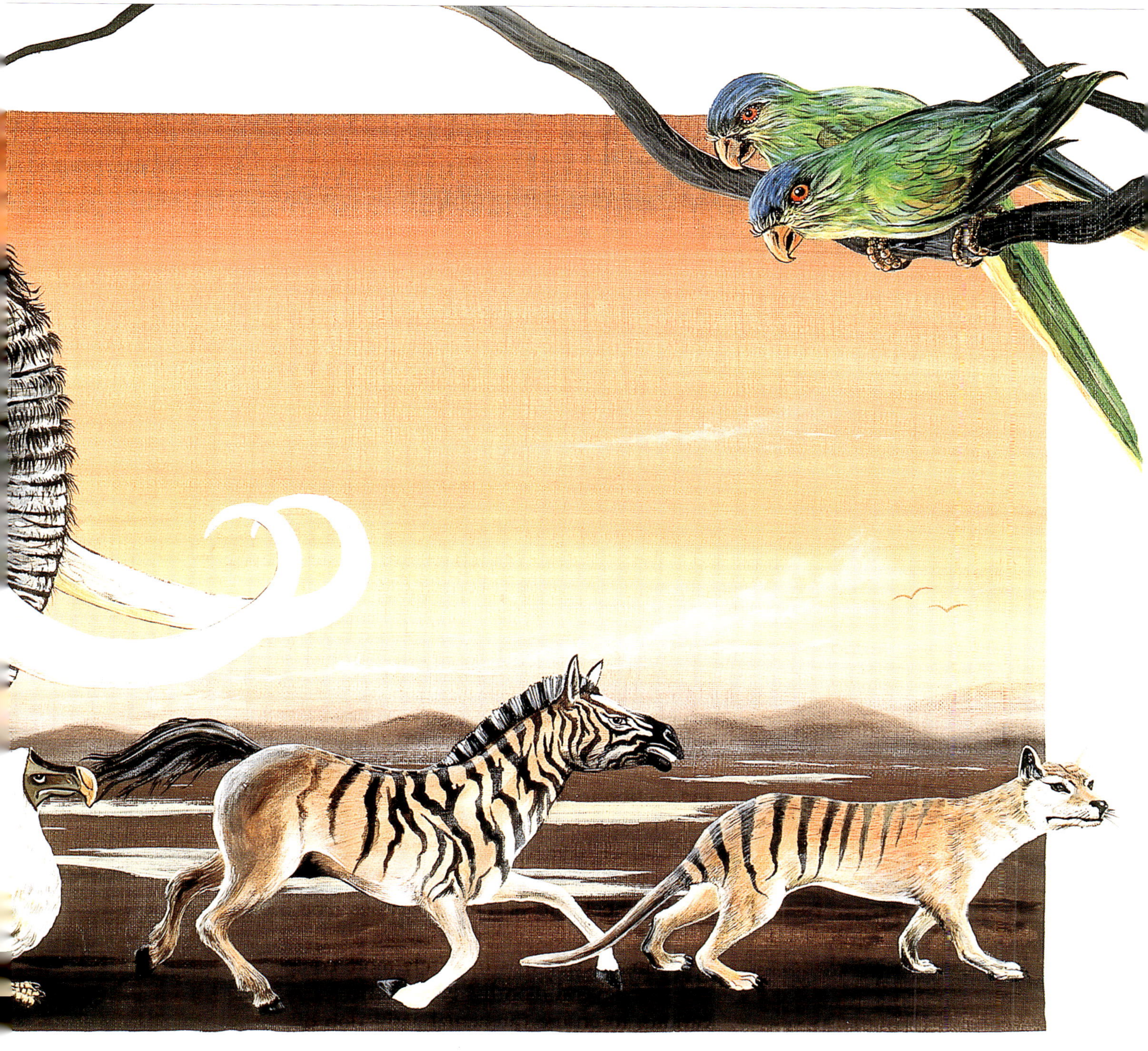

into the fertilised egg of a living organism with morphological similarities to the first species.

An example often quoted by scientists is the quagga, a member of the Equidae family related to both the horse and the zebra, that disappeared from the African continent, on which it had been confined, hardly more than a century ago. The distance between a quagga and a zebra is certainly shorter than that separating thylacine from the wolf-dog or hyena, by the fact that the thylacine belongs to the order of marsupials, whereas the hyena does not, which singularly complicates the problem.

Mummified quaggas exist in various museums and the operation aimed at extracting DNA from their tissue was successfully achieved as early as 1984. Let's then suppose, to start with, that we have managed to identify the gene responsible for the zebra's stripes, a feat we can reasonably consider possible. The first phase, in the laboratory, will be to seek out similar sequences in the quagga DNA, in order to "recover" the corresponding gene within the whole molecular mass, by hybridisation, which means identifying, base by base, the complete DNA sequence that we know (the zebra's) and the one we wish to recover (the quagga's). This operation would not only enable the extraction of the gene to be replaced, but also to define it in its entirety.

Previous spread: The long and dismal list of some famous extinct animals:
- the Mammoth (Europe)
- the Moa (New Zealand)
- the Quagga (Africa)
- the Meiolania Tortoise (Pacific)
- the Dodo (Indian Ocean)
- the Huia (New Zealand)
- the Great Pengoin (Iceland)
- the Diadem Parrot (New Caledonia)
- the Thylacine (Continental Australia and Tasmania).

Then comes the substitution phase. This means, to start with, introducing the given quagga gene into the cell to be modified. Biologists use the name transfection to describe this operation. For the circumstance, the sequence that corresponds to the exogenous gene is provided with catalyst sites by way of restriction enzymes whose role, as we have seen, is to cut the DNA in specific places. This would be the case for the gene that is responsible for the zebra's stripes. In doing this, we achieve what biologists call the "knock out" of the endogenous gene.

This is the theory, but in practice it is infinitely more complicated. Each of the phases described can indeed take several months! For it is necessary, at the end of each phase, to carry out verifications, to ensure the actual activation of the transfected gene and analyse the protein whose synthesis it commanded. In terms of overall research time, all these operations can take from five to six years, monopolising all the attention of a team of three researchers, on condition that each of the described phases occurs successfully…

And supposing that everything indeed goes according to plan, we would only ever have been able to create a false quagga, or in other words a zebra with an aberration in its

CHAPTER ELEVEN — THE IMPOSSIBLE DREAM

fur. As for a wolf-dog or hyena similarly dressed up as a thylacine, they would obviously be without the marsupial pouch.

If the gene responsible for the thylacine's stripes were to be identified one day, from the remains of a specimen, a «transfection» could be attempted to disguise a wolf hound as a thylacine!

Some have nonetheless attacked the problem with great conviction and constructed a variety of theories. Spectators of the movie "*Jurassic Park*" will recall that according to the professor-come-apprentice-sorcerer, it was possible to extract DNA of ancient creatures from the blood of mosquitoes fossilised in amber, the latter having fed on the blood of extinct monsters… and to assert that a DNA molecule was a negative, a living thing, the recreation of the aforementioned animals was nothing more than a perfectly achievable genetic manipulation.

An all too convenient shortcut… For DNA, in chemical terms, constitutes a well defined entity, albeit inert, whereas a cell is a living thing of extreme complexity, in particular because of its structural diversity and the arrangement of the constituents of its nucleus.

Moving from the DNA stage to that of a cell would therefore equate to the move from the inanimate to the living, which, to all intent and purpose, means that this mind-boggling phase is and will without doubt always be beyond human comprehension.

Furthermore, the most eminent biologists humbly admit that even from totally pure DNA that they would extract from their very own tissue, they have absolutely no idea how to create an egg or an embryo, in other words a diploid cell born of the fusion of a male gamete and female gamete, even though the structure of these is well known to them, as is the chemical nature of their constituents.

Let us nonetheless explore pure fiction, and assume that we could one day totally manufacture a living somatic cell, that of our thylacine in particular, using an entirely reconstituted genome. Up until recently, some researchers considered the following scenario with a view to producing such a cell: another egg, sterile and mature, belonging to a species that is morphologically close to thylacine, such as the hyena, would be produced using hormonal techniques. The nucleus of this egg would then be destroyed using a fine laser beam, prior to the fusion of the egg with the somatic cell from the thylacine. The artificial zygote obtained would therefore contain all the chromosomes of the thylacine. It would be implanted in the uterus of a hyena and a baby thylacine could be born.

However, it is obvious that the new-born thylacine, at the embryonic stage, would have a very short life expectancy, given that it would be without the marsupial pouch to provide shelter and food. The chances of success would no doubt be greater with a mother belonging to Tasmanian devil species. But even under these circumstances, there would still be the problem of morphological and nutritional incompatibility.

Right: **The different phases of a hypothetical cloning.**

Professor Wilmut and his collaborators at the Roslin Institute in Edinborough have recently proven to the world that such a scenario is now reality. Their "miracle" consisted of provoking, using a totally assexual method, the birth of the genetic copy (commonly called a clone) of an adult ewe.

The somatic cells of the latter, in this instance taken from the epithelium of its mammary glands, and consequently differenciated[1], were held for five days in a state of quiescence (this is normally a prolonged phase of rest, at the end of mitosis), close to cellular death, through the deprivation of serum in the culture. At the same time, another oocyte, this one fertilised by the natural sexual process, was taken from a second ewe. It underwent enucleation by microdissection and irradiation using ultra-violet rays.

The next phase, by far the most delicate, encompassed the transfer of the cell nucleus that had been placed in culture, into the cytoplasmic environment of the enucleated generalised cell. In such a process, the female gamete only intervenes by way of the reprogramming capacity of the receiving cytoplasmic system, this having been activated. The authors suggest that it is the drastic treatment inflicted on the transplanted nucleus that rendered it accessible to the reprogramming factors of the oocyte.

Assuming that a last female Thylacine specimen is miraculously discovered

Female clone of the original Thylacine

Generalised cell (egg obtained from a pair of Tasmanian Devils)

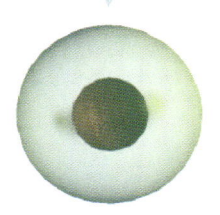

Extraction of a somatic (differenciated) cell from the mammary gland

Prolonged cultivation of this cell

Implantation of this egg into the uterus of a surrogate female devil

Destruction of the nucleus of this cell

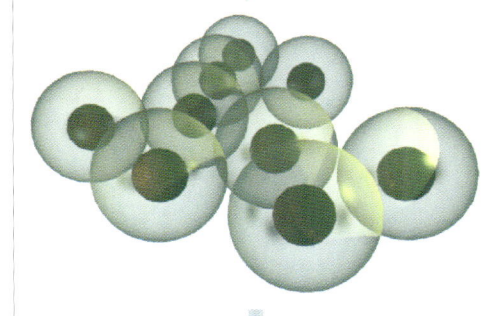

Selection of one of the cells

Transfer of the nucleus into the enucleated egg

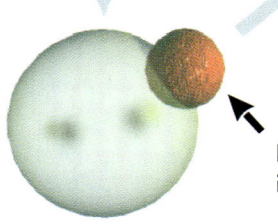

Removal of its nucleus

Enucleated cell

However, in reality, the manipulations were extremely complex. No less than 277 attempts were required in order to create a root-cell and make it develop in the uterus of a third ewe, which, for the circumstance, played the part of a surrogate mother. In this manner, the now famous Dolly finally saw the light of day.

Cloning of mammals had already been carried out successfully before this world first, but always using embryonic cells, with genetic engineers stating that a cell that had reached the limits of differentiation could not, under any circumstances, return to a generalised state. The Scottish scientists have therefore brushed aside this assumption, and opened up exciting perspectives in medical terms.

Dolly's birth, however, has led to some far too hasty conclusions on the part of certain journalists seeking sensational headlines and all too keen to tackle ethical questions, as if one were on the verge of being able to clone human beings at one's leisure. In the first instance, it must not be forgotten that a clone will never be a perfect copy of its model, in so far as the cytoplasm of the cell produced during fertilisation, which could be described as traditional, is preserved. Furthermore, the invariant nature of this environment introduces differentiation factors between the root-individual and its replica, due to the existence of specific organelles, mainly mitochondria.

On top of this, without taking into consideration that the percentage of embryoes obtained by way of the manipulations mentioned above remains very low in the current state of knowledge, no indications as to the longevity of Dolly and her cloned sisters, nor their fertility, have been forthcoming. This last aspect of the problem will however be of prime importance once the same technique is used to produce male animals.

Coming back to the thylacine, the method used by a genetic engineer who manages to create from scratch a somatic cell, using genetic material taken from the remains of an animal - and we have already stated that such an hypothesis is totally utopic - would differ greatly from the material used by the conceivers of Dolly. For Dolly is a product of manipulations carried out on live animals born of natural processes. On of the key points in genetics is indeed that embryonnic development in mammals is only possible using gametes, or haploid sex cells, which only contain one set of chromosomes, and not from a single diploid somatic cell, such as the one that would have been created.

Accordingly, even if one were successful in provoking embryonnic development using such a cell, it could only be abortive because achieving morphogenesis requires half the genome of the cell to be of male origin and the other half of female origin. Indeed, each haploid cell, at the time of fertilisation, contributes specific factors that are necessary for cellular development and the destiny of the newly created cell. It is in this manner, to take but one example, that the mitochondria mentioned above - the organelles that are fundamental to cellular respiration - all come from the mother. Also, different genes are activated at different times, according to whether they come from the male's chromosomes or from those of the female. But, in the case of the diploid genome, both sets of chromosomes would come from the remains of a single animal of a given sex.

Hence, it becomes apparent that the total reconstruction of an extinct species is all myth. Without doubt, it would be less far-fetched and not as pretentious on the part of man to envisage one day reaching and exploring another galaxy, than to master the process of recreation of an extinct higher organism, based on the knowledge we would have of its genome.

Some will conclude that this is well and truly proof of the existence of God. As for agnostics, they will undoubtedly prefer to believe in the omnipotence of the human mind and will remark that molecular biology has led to so many victories, in less than half a century, that what appears today to be inconceivable and inaccessible to the human mind, will in fact be possible in the future, through the progress of that god we call Science.

To each his own "religion". In the meantime, one thing is sure - for all those who are about to close this book - there is absolutely no chance they will one day be able to contemplate a living thylacine produced by genetic manipulation.

1- The initial mitosis (cellular divisions) produce cells which are capable of becoming any of the organism's cells. On average, it is only from the sixteenth mitosis that differentiation begins, giving each cell of the future organism its specificity (blood, bone, tissue, etc.). The generalised cells enable the creation of clones by transplantation into artificially created embryonal envelopes. The experiment has been attempted with success on calf foetuses, but not as yet on human foetuses, an international convention (that only the Republic of Singapore has not ratified) prohibiting all manipulations on these.

Tables to Chapter Seven

Table 7.1 - Sheep lost by predation, Woolnorth, May 1839 to September 1850

Year	Month	Sheep Loss	Predator
1839	May	1	Thylacine
1840	April	1	Thylacine
	May	1	Thylacine
	Sept.	2	Thylacine
1841	March	2	"Vermin"
	July	2	"Vermin"
	Aug.	4	Thylacine
1843	March	4	Dogs
	April	21	Dogs
	June[1]	69	Dogs
		16	Thylacine
	July	24	"Vermin"
	Aug.[2]	47	"Vermin"
		130	Dogs
	Sept.	19	"Vermin"
	Oct.	65	"Vermin"
	Nov.	66	"Vermin"
1844	Feb.	3	"Vermin"
	March	19	"Vermin"
	April	2	"Vermin"
1845	April	23	"Vermin"
	May	35	"Vermin"
	June	10	"Vermin"
	July	28	"Vermin"
	Oct.	7	"Vermin"
	Nov.	38	Thylacine
1846	April	13	Thylacine
	July	14	"Vermin"
	Oct.	36	"Vermin"
1847	Jan.	18	"Vermin"
	Feb.	8	Thylacine and dogs
	March	10	"Vermin"
	April	7	"Vermin"
	May	6	"Vermin"
	June[3]	54	Dogs
	Sept.	67	Dogs
	Dec.	52	Dogs
1848	March	30	Dogs
	June	139	Dogs
	Sept.	94	Dogs
1849	June	305	"Vermin"
	Sept.	216	"Vermin" and weather
1850	Sept.	42	"Vermin"

1- The 69 killed by dogs occurred during a slaughter of 64 Leicester lambs.
2- Of the 130 killed by dogs, no fewer than 78 were driven into the sea.
   From June 1847 the stock return seems to have been sent in quarterly.
3- Source: Compiled from the monthly stock returns of the Van Diemen's Land Company for the Woolnorth Station.

Table 7.2 - Thylacines presented annually for government bounty, 1888-1912

Year	Adults	Juveniles	Total
1888	72	9	81
1889	109	4	113
1890	126	2	128
1891	87	3	90
1892	106	6	112
1893	103	4	107
1894	100	5	105
1895	104	5	109
1896	119	2	121
1897	107	13	120
1898	106	2	108
1899	132	11	143
1900	138	15	153
1901	140	11	151
1902	105	14	119
1903	92	4	96
1904	82	16	98
1905	99	12	111
1906	54(30)	4(3)	58
1907	42(19)	0	42
1908	15	2	17
1909	2	0	2
1910	0	0	0
1911	0	0	0
1912	0	0	0
	2040	144	2184

Note: The numbers for 1906 and 1907 were calculated from Treasury reports, the numbers in brackets being the totals for which bounty was claimed through the Lands Department for the first six months of the year. The Lands Department account books for the last six months are missing.

Table 7.3 - Hunters and their catches of Tasmanian tigers in the 6 regions of Tasmania. Arranged in 5-year groups.

Year	Number of Hunters	Catch	Mean Catch
EAST COAST			
1888-1892	8	106	13.25
1893-1897	13	66	5.07
1898-1902	31	64	2.06
1903-1907	34	53	1.56
TOTAL	86	289	3.47
EASTERN TIERS			
1888-1892	52	97	1.86
1893-1897	38	109	2.88
1898-1902	32	119	3.72

Table 7.3 - Continued

Year	Number of Hunters	Catch	Mean Catch
1903-1907	20	63	2.73
TOTAL			
WESTERN TIERS			
1888-1892	17	120	7.05
1893-1897	26	161	6.19
1898-1902	33	172	5.21
1903-1907	20	98	4.90
TOTAL	96	551	5.74
CENTRAL HIGHLANDS			
1888-1892	23	77	3.35
1893-1897	19	64	3.37
1898-1902	9	107	11.98
1903-1907	3	33	11.0
TOTAL	54	281	5.20
NORTH-WEST			
1888-1892	7	18	2.57
1893-1897	11	34	3.09
1898-1902	52	72	1.38
1903-1907	16	19	1.19
TOTAL	86	143	1.66
WEST COAST			
1888-1892	0	0	0
1893-1897	6	6	1.0
1898-1902	1	3	3.0
1903-1907	3	3	1.0
TOTAL	10	12	1.2

Table 7.4 - Sheep losses on various runs and paddocks at Woolnorth, 1888-95

	1888	1889	1890	1891	1892	1893	1894	1895	TOTAL
Mt Cameron	27	34	18	24	24	17	26	27	197
Forest	0	27	21	0	16	7	11	32	114
Welcome Heath	0	2	2	0	6	0	2	5	17
Three Sticks	0	5	7	18	10	10	7	23	80
Studland Bay	6	13	4	2	2	0	3	6	36
"Paddocks"	21	7	3	13	3	18[1]	10	5	80
Victory Run	0	0	0	15	0	0	0	0	15
	54	88	55	72	61	52	59	98	539

1- No less than 10 of these sheep were poisoned when a new cure for footrot was used.

Table 7.5 - Tasmanian tigers caught, incidents reported, Woolnorth, 1874-1911

Locality	Killed	Incidents	Description
Mt Cameron	2(9)	2	coastal
Studland Bay	3(8)	24	coastal
Three Sticks	2	13	coastal
Valley Bay	4	0	coastal
Forest	1	24	inland
Mc Cabe's	3	0	homestead
Spink's	1	0	homestead
Bullock	1	0	homestead
No details	7	—	—
Green Point	—	3	coastal
Welcome Heath	—	13	inland
Swan Bay	—	2	coastal
W. Mile Marsh	—	1	inland
Inlet	—	6	coastal
Bluff	—	3	coastal
Park Paddock	1	—	coastal
TOTAL	25 (17)	91	
	(29 coastal/6 inland/7?)	(53 coastal/38 inland)	

Note: Bracket refers to Wainwright's catch (see footnote[3], Table 7.6).

Table 7.6 - Details of tiger hunts, Woolnorth, 1874-1914

1874	21 Aug.	One tiger killed.
1875	No comments.	
1876	4 March	Tracked a tiger at Studland Bay.
1877	No comments	
1878	13 Sept.	Trying to shift a tiger up the coast from Studland Bay.
1879	14 July	Trying to shift a tiger at Studland Bay.
1880	6 July	Chased a tiger from Studland Bay run.
1881	No comments	
1882	No comments	Other than collected skins and hides from the Mount.
1883	No comments	
1884	No comments	
1885	No comments	
1886	26 June	Snares set at Green Point[1].
1887	3 Jan.	To Green Point Snares.
	31 Jan.	To Green Point Snares[2].
	8 April	All hands chasing tiger from Forest.
	23 June	To Studland Bay to shift a tiger.
	18 Aug.	Tiger in the Forest.
	19-20 Aug.	Snares set in the Forest.
	23 Aug.	One tiger located at the Mount.
	24 Aug.	Forest snares empty.

Table 7.6 - Continued

		Also empty on 30 Aug., 3, 7, 10, 15 and 28 Sept.
	5 Oct.	Tiger at Studland Bay.
	7 Oct.	Forest snares empty. Also empty on 11 Oct.
	10 Nov.	Mount snares empty. Also empty on 15 Nov.
1888	14 March	Tiger man came home.
	23 May	Tiger man at Arthur River, snares not looked after.
	15 June	Trying to shift tiger out of Forest.
	16 June	Trying to shift tiger out of Forest.
	21 June	Looking after tiger on Swan Bay run.
	28 June	All hands at Studland Bay trying to shift tiger. Tracked him.
1889	26 Aug.	Tom went to the Mount to look after a tiger with his dogs.
	2 Nov.	Sent some men to hunt tiger out of Studland Bay run.
	11 Nov.	All hands in a.m. hunting a tiger out in the Forest.
		Set snares for a tiger on Saltwater Creek fence.
1890	6 Feb.	Tracks seen in Forest.
1891	1 Aug.	Laid poison at Harcus for hunters' dogs.
1892	26 Aug.	Two men to Studland Bay to shift tiger.
1893		No comments
1894	9 April	Davey caught a tiger in Park Paddock.
	28 April	Went to Studland Bay to shift a tiger.
	15 June	Some of the men went to Studland Bay to shift a tiger.
1895	24 June	Tracked a tiger on Welcome Heath.
		Went with dogs in the afternoon to try to shift him.
	25 June	Looking after tiger on Welcome Heath.
1896		Diary no longer in existence.
1897	25 Jan.	Shifting tiger off Welcome by burning the scrub.
	30 April	Track seen on Welcome Heath.
	2 May	Shifting Welcome Heath tiger.
	5 May	Shifting Welcome Heath tiger.
	8 May	Hunting tiger in the Forest.
	9 May	Hunting tiger in Forest.
	10 May	Hunting tiger in Forest.
	22 May	Hunting tiger on heath.
	30 May	Hunting tiger on heath.
	3 June	Tiger seen in the Forest.
	4 June	Tiger chase in Forest, unsuccessful.
	5 June	Tiger chase in Forest.
	17 July	Tiger on Studland Bay and Three Sticks runs.
	5 Aug.	Tiger at work on Studland Bay run.
	6 Aug.	Went to shift tiger.

Table 7.6 - Continued

	14 Aug.	Still looking for Three Sticks run tiger.
	28 Aug.	Still looking for Three Sticks run tiger.
	7 Sept.	Still looking for Studland Bay tiger.
1898	20 Feb.	One tiger caught. No locality given.
	20 July	One tiger caught, McCabe's Paddock.
	31 Dec.	Snaring in the Forest.
1899	3 July	Saw two tigers at Swan Bay.
	6 July	Caught two tigers in Forest and Three Sticks.
	22 July	Tiger scaring on Three Sticks and Studland Bay.
	23 Nov.	One tiger caught, probably at the Mount.[3]
1900	24 Jan.	One tiger caught, locality not stated.
	8 Feb.	Tiger scaring at Three Sticks.
	10 Feb.	Tiger in snares, no locality.
	12 Feb.	One tiger caught, no locality.
	13 Feb.	One tiger caught, no locality.
	25 March	One tiger caught, no locality.
	9 April	All hands tiger scaring, Three Sticks to Studland Bay.
	23 May	One tiger caught at Bullock Paddock.
	13 June	Two tigers caught, probably at the Mount.[3]
	14 June	One tiger caught.[3]
	5 July	All hands tiger scaring on Three Sticks and Studland Bay.
	9 July	One tiger caught, no locality.[3]
	19 July	One tiger caught, the Mount beach.
	12 Aug.	Tiger scaring in Forest.
	13 Aug.	Tiger scaring on Three Sticks and Studland Bay.
	18 Aug.	One tiger caught, no locality.[3]
	19 Aug.	One tiger caught, no locality.[3]
	2 Sept.	Tiger scaring at Studland Bay.
	6 Sept.	Two tigers caught, Valley Bay.
	9 Sept.	Three tigers caught, no locality.[3]
	10 Sept.	One tiger caught, no locality.[3]
1901	8 Jan.	Tiger seen in Western Mile Marsh.
	9 Jan.	Snares set on Welcome Heath and in Forest.
	11 Feb.	Tiger scaring at Forest and Welcome Heath.
	29 April	Tiger scaring in Forest.
	2 May	Tiger snaring at Studland Bay. One tiger caught, no locality.[3]
	16 May	Tiger snaring on Welcome Heath, Inlet and Forest.
	18 May	Tiger scaring and snaring, Forest and Inlet.
	25 May	Snaring in Forest.
	30 May	Scaring and snaring in Forest and Inlet.

Table 7.6 - Continued

	1 June	Snaring at Three Sticks, Bluff, Welcome Heath and Inlet.
	28 June	Two tigers caught, Studland Bay.[3]
	11 July	Scaring tigers in Forest and Inlet.
	13 July	Snaring on Three Sticks and Bluff.
	16 July	Scaring tigers in Forest and Inlet.
	17 July	One tiger 'dogged' on Three Sticks. Scaring and snaring at Forest and Inlet.
	18 July	One tiger in Spink's Paddock and two in Valley Bay snares.
	19 July	One tiger caught in McCabe's Paddock.
	20 July	One tiger caught in McCabe's Paddock.
	31 Dec.	Snaring in the Forest.
1902	10 March	One tiger caught, no locality.[3]
	28 March	Tiger scaring on Welcome Heath.
	24 May	Tiger scaring on Three Sticks.
	26 May	Tiger scaring and snare setting on Forest.
	10 June	One tiger caught, Studland Bay.[3]
	16 June	Snaring in Forest.
	27 June	One tiger caught, no locality.[3]
1903	20 April	Dog caught in tiger snare.[4]
	30 April	T. Well on Three Sticks run, tiger hunting.
	30 May	Four men on Three Sticks run, tiger hunting.
	15 Sept.	Two tigers caught round lambs and ewes.[5]
1904	14 May	All hands to tiger hunt.
1905	19 May	Snares set on Mt Cameron.
	16 June	Snares on Mount Cameron examined.[2]
	14 July	Snares examined again.[2]
1906	19 May	Snares examined, no locality. Also examined on 23 May.
	21 June	Snares examined.[2]
	30 Sept.	One tiger caught at Studland Bay.
1907	13 Aug.	Snares at Studland bay examined.
1908		No activity reported.
1909	7 July	Snaring at Studland Bay.
	11 July	Snares examined, also on 13 July.
1910	26 May	Studland Bay snares examined.
1911		No further activity reported from this year.

1- It seems that when snares were set at unusual places this was duly recorded. Snares were always set from Mount Cameron to Three Sticks.
2- Note the long gaps between examinations, any animal would certainly die in this time.
3- Caught by Wainwright. These animals were presumably caught at either Mt Cameron or Studland Bay since Wainwright worked in those places and stated that he caught nearly all of his seventeen tigers there.
4- One wonders how many dogs were killed in this fashion. The wild dog population certainly would have suffered casualties, all of which would have died before being removed from the snares.
5- This is the only mention of the presence of thylacine around sheep.

Illustration credits

Cover
Front and back cover   Priscilla WRIGHT, Australian animalist painter
Jacket flap   National Philatelic Collection, Australia Post

Chapter One

10, 12	Gerard Willems - *Hobart, Tasmania*
14	Courtesy of B.H. Arnold (*Collection Eric Guiler - Hobart, Tasmania*)
16, 17, 18/19	Linnean Society of London Archives
21	Sketch from Patrick Duffieux - *New Caledonia*
23	National Zoological Park - Smithsonian Institution, *Washington DC*
24/25	Sketch from Patrick Duffieux - *New Caledonia*
26	R.W. Thomas and the South Australian Field Naturalists' Club

Chapter Two

28	Photo X - By courtesy of Museum and Art Gallery of the Northern Territory
30, 31, 32	Photos D. Lowry - *Linden Park, South Australia*
33	Photo A. Baynes, Western Australian Museum - *Perth, W.A.*
34, 35	Photos X, Western Australian Museum - *Perth, W.A.*
36/37, 38, 39, 40	By courtesy of the Nanga-Ngoona Moora-Joorga Land Council - *Roebourne, W.A.*
41	Private collection Eric Guiler - *Hobart, Tasmania*
42, 43	Photos X - By courtesy of Museum and Art Gallery of the Northern Territory
44	Private collection Eric Guiler - *Hobart, Tasmania*

TASMANIAN TIGER — A LESSON TO BE LEARNT

Chapter Three

46, 48, 49, 50	Private collection Eric Guiler - *Hobart, Tasmania*
51	By courtesy of the Queen Victoria Museum - *Launceston, Tasmania*
52, 53	Private collection Eric Guiler - *Hobart, Tasmania*
54, 55	Sketches from François-Gilles Bachelier - *New Caledonia*
56, 57	Private collection Eric Guiler - *Hobart, Tasmania*

Chapter Four

58	Private collection Eric Guiler - *Hobart, Tasmania*
60, 61, 62/63	Photos P. Dombrovskis - *Hobart, Tasmania*
64	Photo P. Godard - *New Caledonia*
66	Photo P. Dombrovskis - *Hobart, Tasmania*
67	Photo P. Godard - *New Caledonia*
68/69	Photo P. Dombrovskis - *Hobart, Tasmania*
70, 71, 72, 73, 74	Photos P. Godard - *New Caledonia*
75	Private collection Eric Guiler - *Hobart, Tasmania*

Chapter Five

76	Musée de la Marine - *Paris, France*
79 (left)	Service Historique de la Marine - *Vincennes, France*
79 (right)	Photo J.L. Charmet - Académie des Sciences - *Paris, France*
81	Batty Library - *Perth, W.A.*
82	Collection Tasmanian Museum & Art Gallery - *Hobart, Tasmania*
83	Private collection Eric Guiler - *Hobart, Tasmania*
84	Sketch from François-Gilles Bachelier (after the original drawing owned by Eric Guiler) - *New Caledonia*

Chapter Six

86	"*Animal Kingdom - The Class Mammalia*" by Baron Cuvier London, Whittaker, 1827
88	Pourrat Publisher - Paris, France (*Collection Gerard Willems - Hobart, Tasmania*)
89 top	"*Naturgeschichte und Abbildungen der Saügetiere*" - Brodtmanns Lithographische Anstalt - Zürich, H. R. Schinz, 1827 (*Collection Gerard Willems Hobart, Tasmania*)
90/91	Illustration presented to the Linnean Society of London in March 1821 by Mr W. Lister Parker. It is a work by John William Lewin, a painter and drawer of animals, born in England in 1770.
92	Author unknown (*Collection Gerard Willems - Hobart, Tasmania*)
93 top	Illustration by Varin in an unidentified work (*Collection Gerard Willems - Hobart, Tasmania*)
93 below	"*The National History of Marsupialia*" W.H. Lizars - Edinburgh, 1841 (*Collection Gerard Willems - Hobart, Tasmania*)
94 top	Sketch from François-Gilles Bachelier (after the catalogue of the London Zoo Guide) - *New Caledonia*
94 below	The newsletter of the "*Royal Zoological Society of London*"
95	Engraver and publisher unknown (*Collection Gerard Willems - Hobart, Tasmania*)
96/97	Lithograph by H.C. Richter in John Gould's "*The Mammals in Australia*" - Vol.1, 1863 (*Collection Gerard Willems - Hobart, Tasmania*)
98	Delineation and lithograph from a photograph by Voctor A. Prout "*The Tasmanian Tiger, Thylacinus cynocephalus*" - Sydney, NSW - Thomas Richards, Government Printer, 1869 (*Collection Gerard Willems - Hobart, Tasmania*)
99 top	Sketch by R. Kretschner in "*Brehm's Tierleben*", 1865

ILLUSTRATION CREDITS

99 below	"The Class Mammalia" by Baron Cuvier - London, Whittaker, 1877
100/101	"Wombat und Beutelwölfe": Schreiber's Bilderwerke - Tome VI, 1882 (Collection Gerard Willems - Hobart, Tasmania)
102	Sketch by J.J. Mahuteau after the original drawing in "Land, Sea and Sky, or Wonders of Life and Nature. A Description of the Physical Geography and Organic Life of the Earth". Translated from the German of Dr Herman, J. Klein and Dr Thomé, by J. Minshull. With three hundred original illustrations, after London, Ward, Lock & Co, Warwick House - Salisbury Square, E.C., 1884 (Collection Gerard Willems - Hobart, Tasmania)
103 top right	Publisher unknown (Collection H. Moeller)
103 top left	"Das Buch für Alle", 1890 (Collection H. Moeller - Heidelberg, Germany)
103 below	Sketch by F. Specht in "Emu, von Beutelwölfen verfolgt" - Das Buch für Alle, 1890 (Collection H. Moeller - Heidelberg, Germany)
104 top	"Cassell's Concise Natural History", Cassell - London - Also appears in Louis Figuier (undated) "Mammalia, their various forms and habits" - same publisher (Collection H. Moeller)
104 below	Sketch by Ferdinand Onjonscoski in "Beutelwolf ein Schnabeltier überfallend", 1892
105	"The Royal Natural History" by Richard Lydekker - London, 1894
106/107	Engraver and publisher unknown (Collection Gerard Willems - Hobart, Tasmania)
108	"Guide to the Galleries of Mammalia in the British Museum" - Natural History p.114 British Museum - London, 1902
109	"Wildlife of the World", Vol 3 p.219 - Warne - London, 1915

Chapter Seven

110, 113	Private collection Eric Guiler - Hobart, Tasmania
114/115	"Neuester Orbis Pictus" or "The World in Pictures for Pious Children" by Friedrich Heller and Walter Benjamin, 1838
118	Private collection Eric Guiler - Hobart, Tasmania
119	Photo Nature Focus - The Australian Museum - Sydney, NSW
120, 123	Private collection Eric Guiler - Hobart, Tasmania
124	AOT photo
126	Private collection Eric Guiler - Hobart, Tasmania
129	Collection Dr J. Calaby - Canberra, A.C.T.
130/131	Tasmanian Museum and Art Gallery - Hobart, Tasmania
132	Private collection Eric Guiler - Hobart, Tasmania
134	Photo P. Godard - Queen Victoria Museum - Launceston, Tasmania
135, 138, 139, 140	Private collection Eric Guiler - Hobart, Tasmania

Chapter Eight

142	Illustrated Sydney News, April 1867 (Collection Gerard Willems)
145	Private collection Eric Guiler - Hobart, Tasmania
146, 148/149	Sketches from Patrick Duffieux - New Caledonia
151	Photo courtesy of J. Binks - Devonport, Tasmania
152/153, 155, 156	Sketches from Patrick Duffieux - New Caledonia
158	Tasmanian Museum and Art Gallery - Hobart, Tasmania
159	Courtesy of Paul Donovan - Dunedin, N.Z.
160/161, 162/163	Sketches from Patrick Duffieux - New Caledonia
164	Sketch from Patrick Duffieux - New Caledonia
165	Private collection Eric Guiler - Hobart, Tasmania

Chapter Nine

166	Private collection Eric Guiler - Hobart, Tasmania
168	Photo courtesy of G. Roberts - Launceston, Tasmania

169, 170/171	Private collection Eric Guiler - *Hobart, Tasmania*
172	Smithsonian Institution - *Washington DC*
173 top & below	Private collection Eric Guiler - *Hobart, Tasmania*
174, 175	Private collection Eric Guiler - *Hobart, Tasmania*
176	Archives Office of Tasmania

Chapter Ten

178, 180, 181, 182	Private collection Eric Guiler - *Hobart, Tasmania*
183	Photo Courtesy of P. Meyer
188, 189	Private collection Eric Guiler - *Hobart, Tasmania*
191	Courtesy of the Launceston Examiner
192, 193, 194, 195	Private collection Eric Guiler - *Hobart, Tasmania*
197, 198, 200, 201	Private collection Eric Guiler - *Hobart, Tasmania*
203, 204, 205	Private collection Eric Guiler - *Hobart, Tasmania*
206	Photo J. Barnett
208	Photo Brozek (Gamma distribution) - Austral International Press - *Sydney, NSW*
209, 210	Private collection Eric Guiler - *Hobart, Tasmania*
213	Photo courtesy of N. Mooney
214	Department of Land Parks and Wildlife - *Tasmania*

Chapter Eleven

216	Tasmanian Museum and Art Gallery - *Hobart, Tasmania*
220/221, 223	Sketches from Patrick Duffieux - *New Caledonia*
225	Drawing by Maxence Pindon - *New Caledonia*

Bibliography

ALLEN, H.	1958	Letter to Eric Guiler
ALLPORT, M.	1868 a	Remarks on Mr. Krefft's notes on the Fauna Tasmania. *Pap. Proc. Roy. Soc. Tasm.* 1868, 33-6
ALLPORT, M.	1868b	Notes on the Fauna of Tasmania. Appendix to *Pap. Proc.Roy. Soc. Tasm.* 1868.
ANDERSON, H. H.	1905	*A Geography of Tasmania.* Sydney: William Brooks.
ANGAS, G.F.	1862	*Narrative of Australia: a popular account.* London: Society for Promotion of Christian Knowledge.
Animals & Birds Protection Board	1952	Minutes. Aot, AA592.
Animals & Birds Protection Board	1956	Minutes, 13 Nov., 8 Apr. & 4 Dec. 1957- 24 Sept. 1958. AOT AA592
ANONYMOUS	1980	Suche nach dem Beutelwolf mit automatischen Kameras. *Frankfurter Allgemeine Zeitung*, 25 June.
ARCHER, M.	1971	A re-evaluation of the Fromm's Landing thylacine tooth. *Proc. Roy. Soc. Vic.* vol. 84, 229-34.
ARCHER, M.	1974	New information about the Quartenary distribution of the thylacine (Marsupialia, Thylacinidae) in Australia. *J. Roy. Soc. Western Aust.* vol. 57, 43-50.
ARCHER, M.	1976a	The Dasyurid dentition and its relationships to that of Didelphids, Thylacinids and Borhyaenids (Marsupicarnivora) and the Peramelids (Peramelina, Marsupialia). *Aust. J. Zool. Supp. Ser.* vol. 39, 1-34.
ARCHER, M.	1976b	The basicranial region of marsupicarnivores (Marsupialia), interrelationships of carnivorous marsupials, and the affinities of the insectivorous marsupial peramelids. *Zool. J. Linn. Soc. Lond.* vol. 59 (3), 217-322.
ARCHER, M.	1978	The status of Australian dasyurids, thylacinids and myrmecobiids. In M.J. Tyler (ed.) *The Status of Endangered Australian Wildlife*, pp. 29-43. Roy. Zool. Soc. S. Aust.
ARCHER, M.	1982	A review of Miocene thylacinids (Thylacinidae: Marsupialia), the phylogenetic position of the Thylacinidae and the problem of apriorism in character analyses. In M. Archer (ed.) *Carnivorous Marsupials*, vol. 2, pp. 445-76. Sydney: Roy. Zool. Soc. NSW.
ARNOLD, B.H.	1994	*Letters of G. P. Harris.* Arden Press, Sorrento.

ASDELL, S.A.	1946	*Patterns of Mammalian Reproduction.* New York: Comstock.
BACKHOUSE, J.	1843	*Account of a visit to the Australian Colonies.* London: Hamilton, Adams.
BARNETT, C.H.	1970	Talocalcaneal movements in mammals. *J. Zool. Lond.* vol.160, 1-7.
BEDDARD, F.E.	1891	On the pouch and brain of the male thylacine. *Proc.Zool. Soc. Lond.* 138-45.
BEDDARD, F.E.	1903	Exhibition of and remarks upon sections of the ovary of the thylacine. *Proc. Zool. Soc. Lond.* 116.
BELL, E.A.	1965	*An Historic Centenary.* Hobart: Mercury Press.
BELL, E.A.	1967	Article in *Australian Women's Weekly,* 10 May, p. 10.
BENSLEY, B.A.	1903	On the evolution of the Australian Marsupialia, with remarks on the relationships of the marsupials in general. *Trans. Linn. Soc. Lond. (Zool.)* vol. 9.
BERESFORD, Q. & BAILEY, G.	1981	*Search for the Tasmanian tiger.* Blubberhead, Hobart.
BINKS, C.J.	1980	*Explorers of Western Tasmania.* Launceston: Mary Fisher Bookshop.
BLAINEY, G.	1966	*The Tyranny of Distance.* Melbourne: Sun books.
BOARDMAN, W.	1945	Some points of the external morphology of the pouch young of the marsupial Thylacinus cynocephalus. *Proc. Linn. Soc. NSW* vol. 70, 1-8.
BOWLER, J.H., HOPE, G.S., JENNINGS, J.N., SINGH, G. & WALKER, D.	1976	Late Quaternary climates of Australia and New Guinea. *Quaternary Res.* vol. 6, 359-94.
BRANDER,B., HARRELL,M.A., & HOLTHOUSE, H.	1968	"Australia". *Washington: Nat. Geog. Soc.*
BRANDL, E.	1973	*Australian Aboriginal Paintings in Western and Central Arnhem Land.* Canberra: Aust. Inst. Aboriginal Studies.
BRETON, W.H.	1834	*Excursions in New South Wales, Western Australia and Van Diemen's Land.* London: R. Bentley.
BRETON, W.H.	1846	*Excursion to the Western Range.* Tasm. J. vol. 2, 121-41.
BRETON, W.H.	1847	Description of a large specimen of *Thylacinus harrisii*.Tasm. J. vol. 3, 125-6.
BRIGGS, A.L.	1961	Letter to Eric Guiler.
BROWN, B.	1972	The Tasmanian tiger. *Skyline,* 29-31.
BROWN, R.	1973	Has the thylacine really vanished? *Animals* vol. 15 (9), 416-9.
BROWN, R.	1983	Is there hope for our tiger? The Tasmanian Mail 16 August, p.8.
BUCHMANN, O.L.K. & GUILER, E.R.	1977	*Behaviour and ecology of the Tasmanian devil Sarcophilus harrisii.* In Stonehouse & Gilmore (eds) *Biology of Marsupials,* pp. 155-68. London: Macmillan.
BUCKINGHAM, G.	1980	Letter to Eric Guiler.
BUCKLY, P.	1988	*Robbins Island Saga.* Smithton.
BUNCE, D.	1857	*Australasiatic Reminiscences.* Geelong: J. Brown.
BURBURY, F.	1953	Letter to Eric Guiler.
CALABY, J.H. & CÄSAR, C.	1971	Man, fauna and climate in Aboriginal Australia. In J.D. Mulvaney & J. Golson (eds) *Aboriginal Man and Environment in Australia.* Canberra: ANU Press.
CÄSAR, C.	1996	Der Beutelwolf. Leben und Sterben einer Tierart. *Zoologisches Museum der Universität - Zürich*
CAVE, A.J.E.	1968	Mammalian olecranon epiphyses. *J. zool. Lond.* vol. 156, 333-50.
COLEMAN, R.	1976	There's a strange, strange beastie out there. *Melbourne Herald* 20 March, p. 29.
COLLINS, L.R.	1973	*Monotremes and Marsupials.* Washington: Smithsonian Institute.
COOK, D.L.	1963	*Thylacinus and Sarcophilus* from the Nullarbor Plain. Western Aust. Nat. vol. 9, 47-8.
CORRESPONDENT.	1924	Article on the killing of a tiger at Waratah by C. Penny. *Weekly Courrier* 17 January, pp. 26, 46.
CRISP, E.	1855	On some points relating to the anatomy of the Tasmanian Wolf (*Thylacinus*) and of the Cape Hunting Dog. (*Lycaon pictus*). Proc. zool. Soc. Lond. 1855, 188-91.
CROWTHER, W.L.	1883	Correspondence. Proc. zool. Soc. Lond. 1883, 252.
CUNNINGHAM, D.J.	1882	Report on some points in the anatomy of the thylacine (*Thylacinus cynocephalus*), cuscus (*Phalangista maculata*) nd phascogale (*Phascogale calura*) collected during the voyage of H.M.S. Challenger in the years 1873-6. *Challenger Rept Zool.* vol. V (16), 1-192.

DAVIES, J.	1886	Parliamentary reports in the *Hobart Mercury*, 5 November.
DAVIES, J.L.	1965	*Atlas of Tasmania*. Hobart: Govt Printer.
DOUGLAS, A.M.	1990	The thylacine: A case for current existence on mainland Australia. *Cryptozoologie*, 9, 13-25.
DUNMORE, J.	1983	*French explorers in the Pacific*. vol. 2, 19th Century. Soc. Nouv. des Editions du Pacifique. Tahiti.
DUNNET, G.M. & MARDON, D.K.	1974	A monograph of Australian Fleas (Siphonaptera). Aust. J. zool. Supp. Ser. vol. 30, 1-273.
EASTWOOD, D.	1981	Letter to Eric Guiler.
EVANS, G.W.	1822	*Description of Van Diemen's Land*. London: Souter.
E WENCE, G.	1961	Article in The Bulletin, 9 September, p. 29.
FITZGERALD, L.	1984	*Java La Grande, Portuguese Discovery of Australia*. The Publishers, Hobart.
FLEAY, D.	1946	On the trail of the marsupial wolf. Vict. Nat. vol. 63, 129-35; 154-59; 174-77.
FLEMING, A.L.	1939	Reports on two expeditions in search of the thylacine. J. Soc. Preserv. Fauna Emp vol.30, 20-5.
FLYNN, T.T.	1914	The Mammalian fauna of Tasmania. Brit. Assoc. Adv. Sci. Handb. 48-53.
FLOWER, S.S.	1931	Contribution to our knowledge of the duration of life of vertebrate animals. Proc. zool. Soc. Lond.145-243.
FLOWER, W.H.	1865	On the commissures of the cerebral hemispheres of the Marsupialia and Monotremata as compared with those of Placental Mammals. Phil. Trans. Roy. Soc. Lond. vol. 55, 633-51.
FLOWER, W.H.	1867	On the development and succession of the teeth in the Marsupialia. Phil. Trans. Roy. Soc. Lond. vol. 155, 631-41.
FOLTELNY, J.G.	1967	Gibt es einen australischen Tiger? *Kosmos* vol. 63, 292-4.
GEOFFROY SAINT-HILAIRE, E	1810	Description de deux espèces du *Dasyurus*. Ann. du Mus. vol. 15, 301-6.
GILL, E.D.	1953	Distribution of the Tasmanian devil, the Tasmanian wolf and the dingo in South East Australia in Quaternary time. Vict. Nat. vol. 70, 86-90.
GILL, E.D.	1964	The age and origin of the Gisborne Cave. Proc. Roy. Soc. Vict. vol. 77, 532-3.
GLAUERT, L.	1914	The Mammoth Cave. Rec. Western Aust. Mus. vol. 1, 244-51.
GODFREY, M.	1984	*Waratah - Pioneer of the West*. Waratah Municipality, Waratah.
GODWIN	1823	*Godwin's Emigrants Guide to Van Diemen's Land*. Sherwood Jones, London.
GOLDIE, A.	1829	Report from Goldie to Curr, 13 March 1829. Van Diemen's Land Co. Papers, State Archives of Tasmania.
GOODRICK, J.	1977	*Life in Old Van Diemen's Land*. Adelaide: Rigby, 220 pp.
GOULD, J.	1863	*Mammals of Australia* vol. 1. London: Taylor & Francis.
GOWLAND, R. & GOWLAND, K	1986	*Trampled Wilderness*. Richmond, Devonport.
GRAVES, K.	1958	The rarest animal in the world. *Walkabout*, 1 May, 15-16.
GREEN, R.H.	1967	Notes on the devil (*Sarcophilus harrisii*) and the quoll (*Dasyurus viverrinus*) in north-eastern Tasmania. Rec. Queen Vict. Mus. vol. 27, 1-12.
GREEN, R.H.	1974	Mammals. In *Biogeography and Ecology in Tasmania*, pp. 376-96. Den Hague.
GRIFFITHS, J.	1973	The thylacine on the Central Plateau. In *The Lake Country*, ed. M. Banks. Royal Soc. Tasm., 119-24.
GUILER, E.R.	1958	Observations on a population of small marsupials in Tasmania. J. Mammal. vol. 39, 44-58.
GUILER, E.R.	1961a	The former distribution and decline of the thylacine. Aust. J. Sci. vol. 23, 207-10.
GUILER, E.R.	1961b	The breeding season of the thylacine. J. Mammal. vol. 42, 396-7.
GUILER, E.R.	1966	In pursuit of the thylacine. Oryx vol. 8, 307-10.
GUILER, E.R.	1967	The fauna of Tasmania. Tasm. Year Bk vol. 1, 58-64.
GUILER, E.R.	1970	Observations on the Tasmanian devil, *Sarcophilus harrisii* (Dasyuridae: Marsupialia). II. Reproduction, and growth of pouch young. Aust. J. Zool. vol. 18, 63-70.
GUILER, E.R.	1985	*Thylacine; The tragedy of the Tasmanian tiger*. Oxford, Melbourne.
GUILER, E.R.	1986	The Beaumaris Zoo in Hobart. Proc. Tasm. Hist. Res. Ass. 33, 121-72.
GUILER, E.R.	1991	*The Tasmanian tiger in pictures*. St. Davids Park Press, Hobart.
GUILER, E.R.	1995	*The Enthusiastic Amateurs*. Nat. Parks Service. Hobart. In Press.

GUILER, E.R. & HEDDLE, R.W.L	1970	Testicular and body temperature in the Tasmanian devil and three other species of marsupials. Comp. Biochem. *Physiology*, 33, 881-91.
GUILER, E.R. & MELDRUM, G.K.	1958	Suspected sheep killing by the thylacine, *Thylacinus cynocephalus* (Harris). Aust. J. Sci. vol. 20, 214.
GUNN, R.C.	1850	Letter to the Zoological Society. *Proc. zool. Soc. Lond.* 1850, 90-1.
GUNN, R.C.	1851	Note in *Pap. & Proc. Roy. Soc. V.D.L.* 9 Jul.
GUNN, R.C.	1852	A list of the mammals indeginous to Tasmania. *Pap. Proc. Roy. Soc. Tasm.* vol. 2, 77-90.
GUNN, R.C.	1863	Extracts from a letter to the Secretary of the Zoological Society. *Proc. zool. Soc.Lond.* 1863, 103-4.
HAITCH, R..	1981	Rare tiger quest. *New York Sunday Times* 29 March, p. 41.
HALLSTROM, E.	1961	Letter to N. Laird. Laird Papers, AOT. NS 1143.
HARRIS, G.P.	1808	Descriptions of two new species of *Didelphis* from Van Diemen's Land. *Trans. Linn. Soc. Lond.* vol. IX, 174.
HAYES, J.	1972	Account of thylacine as related by G. Stevenson. *Examiner Express.* 10 June.
HENDERSEN, J.	1832	Observations on the Colonies of New South Wales and Van Diemen's Land. Calcutta: Baptist Mission Press.
HICKMANN, V.V.	1955	The Tasmanian Tiger. *Etruscan* vol. 5 (2), 8-11.
HILL, J.P.	1900	On the foetal membranes, placentation and parturition of the native cat (*Dasyurus viverrinus*). Anat. Anz. vol. 18, 364-73.
HCC (Hobart City Council)	1922	Minutes-reserves committee, 18 July.
HCC.	1923	Minutes-reserves committee, 19 June, 3 July.
HCC.	1924	Minutes-reserves committee, 19 February.
HCC.	1925	Minutes-reserves committee, 21 July.
HCC.	1930	Minutes-reserves committee, 14 April.
HCC.	1935	Minutes-reserves committee, 13 February, 3 July.
HCC.	1936	Minutes-reserves committee, 16 September.
HCC.	1937	Minutes-reserves committee, 17 February, 3 March, 4 April.
Hobart Mercury	1864	26 November, p. 4.
Hobart Mercury	1874	Advertisement by Jemrack, 16 July.
Hobart Mercury	1979	28 December, p. 2.
Hobart Mercury	1980	Article, 20 March.
Hobart Mercury	1983	News item, 23 September, p. 1.
Hobart Mercury	1984	21 January, p. 1.
Hobart Town Gazette	1823	Article on a thylacine incident. 2 August, p. 2.
HOCKING, G.H.& GUILER, E.R.	1983	The mammals of the Lower Gordon River region, South-West Tasmania. *Aust. Wildl. Res.* vol. 10, 1-23.
Hong Kong Standard	1980	Tasmania Tiger eludes search. 21 September.
HOWLETT, R.M.	1960	A further discovery of *Thylacinus* at Augusta, Western Australia. *Western Australian Nat.* vol. 7, 136.
HULL, H.M.	1871	*Hints to Emigrants intending to proceed to Tasmania.* Hobart Town: Fletcher, 48 pp.
INGRAM, B.S.	1969	Sporomorphs from the desiccated carcases of mammals from Thylacine Hole, Western Australia. *Helictite* vol. 7, 62-6.
IREDALE, T. & TROUGHTON, E. le G.	1934	A checklist of the mammals recorded from Australia. *Aust. Mus. Mem.* vol. VI, 15.
JEFFREYS, C.H.	1820	Geographical and descriptive Delineations of the Island of Van Diemen's Land. London: Richardson.
JOHNSTON, R.M.	1890	*Tasmanian Official record.* Government Printer, Hobart.
JONES, M.	1995	Ph. D. thesis, University of Tas. Hobart.
JONES, R.	1970	*Tasmanian Aborigines and dogs. Mankind.* vol. 7, 256-71.
JORDAN, A.	1987	*Tiger Man.* Wordswork, Tasmania.
KEAST, A.	1982	The thylacine (Thylacinidae: Marsupialia) how good a pursuit carnivore? In M. Archer (ed.) *Carnivorous Marsupials*, pp. 675-84. Sydney: Roy. Zool. Soc. NSW.
KEELING, C.H.	1990	Letter to Eric Guiler. 14 Jan.
KENDRICK, G.W. & PORTER, J.K.	1973	Remains of a thylacine (Marsupialia: Dasyuroidea) and other fauna from caves in the Cape Range, Western Australia. *J. Roy. Soc. Western Aust.* vol. 56, 116-22.

KIRSCH, J.A.W. & ARCHER, M.	1982	Polythetic cladistics, or when parsimony's not enough: the relationships of carnivorous marsupials. In M. Archer (ed.) *Carnivorous Marsupials*, pp. 595-619. Sydney: Royal Zool. Soc. NSW.
KNOPWOOD, R.	1805	*The Diary of the Rev. Robert Knopwood, 1803-38.* Hobart: (1977)Tasmanian Historical Research Association.
KREFFT, G.	1868	Description of a new species of thylacine (*Thylacinus breviceps*). Ann. Mag. Nat. Hist. vol. 4 (ii), 296-7.
LAIRD, N.	1968	Article in *Hobart Mercury*, 7 October.
LAIRD, N.		Laird Papers. AOT, NS 1143.
LARKIN, R.	1978	Article in *Hobart Mercury*, 24 February.
Launceston Examiner	1968	News item, 26 April, p. 1.
LE FEVRE, J.	1953	Letter to Eric Guiler.
LE FEVRE, P.	1953	Letter to Eric Guiler.
LE SOUEF, A.S.	1926	Notes on the habits of certain families of the OrderMarsupialia. Proc. zool. Soc. Lond. 1926, 935-7.
LE SOUEF, A.S. & BURREL, H.	1927	*Wild Animals of Australasia*. London: Harrap.
LESTER, C.	1983	Disease takes tiger toll. *Tasm. Mail.* 30 August, p. 2.
LIGHTON, R.F.	1968	Tasmanian tiger believed sighted. *Sci. News* vol. 93, 569.
LLOYD, G.T.	1862	*Thirty-three Years in Tasmania and Victoria.* London: Houlston & Wright.
LORD, C.L.	1928	Existing Tasmanian marsupials. Pap. Proc. Roy. Soc. Tasm. 1927 (1928), 17-24.
LORD, C.L. & SCOTT, H.H.	1924	*A Synopsis of the Vertebrate Animals of Tasmania.* Hobart: Oldham, Bedcome & Meredith.
LOWRY, D.C. & J.W.J.	1967	Discovery of a thylacine (*Thylacinus cynocephalus*) in a cave near Eucla, Western Australia. *Helictite* vol. 5 (2), 25-9.
LOWRY, J.W.J.	1972	The taxonomic status of small fossil thylacines, (Marsupialia; Thylacinidae), from Western Australia. J. Roy. Soc. Western Aust. vol. 55, 19-29.
LOWRY, J.W.J. & MERRILEES, D.	1969	Age of the desiccated carcase of a thylacine (Marsupialia; Dasyuroidea) from Thylacine Hole, Nullarbor Region, Western Australia. *Helictite* vol. 7 (1), 15-6.
LUCAS, A.H. & LE SOUEF, A.S	1909	*Animals of Australia.* Melbourne: Whitcombe & Tombs.
LUKAS, J.A.	1963	Plan to hold that elusive tiger. *Toronto Globe and Mail* 24 December.
LYCETT, J.	1824	*Views in Australia.* London: J. Souter.
LYNE, A.G.	1959	The systematic and adaptive significance of the vibrissae in the Marsupialia. Proc. zool. Soc. Lond. vol. 133, 79-133.
LYNE, A.G. & Mc MAHON, T.S	1951	Observations of the surface structure of the hairs of Tasmanian Monotremes and Marsupials. Pap. Proc. Roy. Soc. Tasm. 1950 (1951), 71-84.
LYNE, J.	1886	Parliamentary report in the *Hobart Mercury*, 5 November.
LYNE, J.	1887	Parliamentary report in the *Hobart Mercury*, 27 August.
LYON, B.	1972	Letter to Eric Guiler.
MACKIESON, D.	1981	Letter to Eric Guiler.
MACINTOSH, N.W.G.	1975	The origin of the dingo: an enigma. In M.W. Fox (ed.) *The Wild Canids*, pp. 85-106. New York: Van Nostrand.
MALLEY, J.F.	1972	The Report of the Search for the Thylacine. Privately Printed.
MALLEY, J.F.	1973	The thylacine. Proc. Seminar Arthur River to the Pieman River. Smithton.
MARTIN, R.M.	1836	Van Diemen's Land. *In History of Australasia*, pp. 282-85. London: J. Mortimer.
MATTHEWS, L.W.	1958	Letter to Eric Guiler.
MATTINGLEY, E.H.	1946	Thylacine and Thylacoleo. Vict. Nat. vol. 63, 143.
Mc INTYRE, K.	1977	*The Secret Discovery of Australia.*
Mc NAMARA, M.	1983	Article in Hobart Mercury, 6 December, p. 1.
MEAD, I.	1961	Letter to Eric Guiler regarding the identity of 'J.S.'.
MELVILLE, P.	1833	*The Van Diemen's Land Almanak.* Hobart: J. Ross, 32-3.
MEREDITH, L.A.	1881	*Tasmanian Friends and Foes.* London: Marcus Ward.
MERRILEES, D.	1968	Man the destroyer: Late Quaternary changes in the Australian marsupial fauna. J. Roy. Soc. Western Aust.vol. 51, 1-24.
MERRILEES, D.	1970	A check on the radio-carbon dating of dessicated thylacine (marsupial 'wolf') and dingo tissue from Thylacine Hole, Nullarbor region, Western Australia. *Helictite* vol. 8 (2), 39-42.

MESTON, A.L.	1958	The Van Diemen's Land Company, 1825-42. *Rec. Queen Vict. Mus.* vol. 9, 1-62.
MILLIGAN, J.	1853	Remarks on the habits of the wombat, hyaena and certain species of reptiles. *Pap. Proc. Roy. Soc. Tasm.* vol. 2, 310.
MILLIGAN, J.	1859	Vocabulary of dialects of the Aboriginal tribes of Tasmania. *Pap. Proc. Roy. Soc. Tasm.* vol. 3, 239-74.
MITCHELL, P.C.	1916	Further observations on the intestinal tract of mammals. *Proc. zool. Soc. Lond.* 183-252.
MOELLER, H.	1968	Zur Frage der Parallelerscheinungen bei Metatheria und Eutheria. Vergleichende Untersuchungen an Beutelwolf und Wolf. *Z. wiss. zool.* vol. 177 (3/4), 283-392.
MOELLER, H.	1970	Vergleichende Untersuchungen zum Evolutionsgrad der Gehirne grosser Rabbeutler (*Thylacinus, Sarcophilus und Dasyurus*). *Z. f. zoologische Syst. und Evol. Fors.* vol. 8, 69-88.
MOELLER, H.	1993	Beutelwölfe in Zoologischen Garten und Museen. *Zeitshr. des Kölner Zoo* 36 (2) 67-71.
MOELLER, H.	1994	Über den Schauwert des Beutelwolfs, Thylacinus cynocephalus. *Zool. Garten*, NF, 64, 2, 97-109.
MOELLER, H.	1997	Der Beutelwolf. Westarp Wissenschaften. Magdeburg.
MOLLISSON, B.C.	1951	Statements attributed to his grandfather. Queen Vict. Mus. files, quoted in S.J. Smith 1980.
MOONEY, N.J.	1984	Tasmanian tiger sighting casts marsupial in new light. *Aust. Nat. Hist.*, 177-80.
MORRIS, D.	1962	The thylacine re-discovered. *Proc. zool. Soc. Lond.* vol. 138, 4, 668.
MORRISSON, B.	1961	*Perth Daily News*, 18 Aug.
MOSELEY, I.	1968	*Hobart Mercury*, 4 Oct.
MUDIE, R.	1829	The Picture of Australia: exhibiting New Holland, Van Diemen's Land and all the settlements from the first at Sydney to the last at Swan River. London: Whittaker, Treacher.
MULVANEY, J.D.	1969	*The prehistory of Australia*. London: Thames & Hudson.
MUNDAY, B.L. & GREEN, R.H.	1972	Parasites of Tasmanian Fauna. II. Helminthes. *Rec. Queen. Vict. Mus.* vol. 44, 1-15.
NICHOLLS, W.	1960	Letter to Eric Guiler.
OWEN, R.	1837	On the structure of the brain in marsupial animals. *Phil. Trans. Roy. Soc. Lond.* 87-96.
OWEN, R.	1841	*Marsupialia*. Todd's Cyclopaedia of Anatomy vol. 3, 257-81.
OWEN, R.	1842	*Trans. Geol. Soc. Lond.* 6. 2nd. Ser.
OWEN, R.	1843	On the rudimentary marsupial bones in the *Thylacinus*. *Proc. zool. Soc. Lond.* vol. 11, 148-9.
OWEN, R.	1846	On the rudimental marsupials bones in the *Thylacinus*. *Tasm. J.* vol. 2, 447-9.
OWEN, R.	1868	*Anatomy of Vertebrates* vol. III. London: Longmans Green.
OWEN, R.	1877	*Researches on the fossil remains of Australia, with a notice of the extinct marsupials of England.* London: the author.
OXLEY, J.	1810	Account of the settlement at Port Dalrymple. *Hist. Rec. Austr. Ser.* III, 111, 1, 771.
PACKER, H.	1970	Is it the Tasmanian Tiger? *Australian Women's Weekly* February 25.
PADDLE, R.N.	1992	Last resting place of a thylacine. Nature, 360, 1992, 215.
PADDLE, R.N.	1993	Thylacines associated with the Royal Zoological Society of N.S.W. *Austr. Zoolog.* 29, 97-101
PARKER, H.W.	1833	*The rise, progress and presents status of Van Diemen's Land*: London: J. Cross.
PARTRIDGE, J.	1967	A 3,300 year old thylacine (Marsupialia: Thylacinidae) from the Nullarbor Plain, Western Australia. *J. Roy. Soc. Western Aust.* vol. 50, 57-9.
PATERSON, W.	1805	Report in *Sydney Gazette and New South Wales Advertiser* vol. 3, 24 April.
PEARSE, A.M.	1981	Aspects of the biology of *Uropyslla tasmanica*. MSc thesis, University of Tasmania.
PEARSON, J. & de BAVAY, J.M.	1953	The urogenital system of the Dasyurinae and Thylacininae (Marsupialia: Dasyuridae). *Pap. Proc. Roy. Soc. Tasm.* vol. 87, 175-99.
PLOMLEY, N.J.B.	1966	*Friendly Mission*. The Journals of G.A. Robinson. Hobart: Tasmanian Historical Research Association.
PLOMLEY, N.J.B.	1983	*The Baudin Expedition and the Tasmanian Aborigines*. Blubberhead. Hobart.
POCOCK, R.I.	1914	On the facial vibrissae of mammalia. *Proc. zool. Soc. Lond.* 1914, 889-912.
POCOCK, R.I.	1926	The external characters of *Thylacinus, Sarcophilus* and some related marsupials. *Proc. zool. Soc. Lond.* 1926, 1037-84.
Pretoria News.	1981	Search on for tiger with a pouch. 24 October, p. 8.
RANSON, B.H.	1905	Tapeworm cysts (*Dithyridium cynocephali n. sp.*) in the muscles of a marsupial wolf (*Thylacinus cynocephalus*), *Trans. Amer. Micros. Soc.* vol. 27, 31-2.

RENSHAW, G.	1938	The thylacine. *J. Soc. Preserv. Fauna Emp.* vol. 35, 47-9.
RIDE, W.D.L.	1964	A review of Australian fossil marsupials. *J. Roy. Soc. West. Aust.* vol. 47, 97-131.
RIDE, W.D.L.	1970	*A guide to the native animals Australia.* Melbourne: Oxford University Press.
RITCHIE, B.	1961	Letter to Eric Guiler.
ROBERTS, M.G.	1915	The keeping and breeding of Tasmanian devils (*Sarcophilus harrisii*). *Proc. zool Soc Lond.* 1915, 575-81.
ROSS, J.	1830	*The Hobart Town Almanack.* Hobart: J. Ross.
ROTH, H.L.	1899	*The Aborigines of Tasmania.* Hobart: Fuller.
ROUNSEVELL, D.E.	1983	Thylacine. In *The Complete Book of Australian Mammals.* Sydney: Australian Museum.
ROUNSEVELL, D.E. & SMITH, S.J.	1980	Recent alleged sightings of the thylacine in Tasmania. Paper read at the Aust. Mammal Soc. Conf.
SACK, M.	1981	Letter to Eric Guiler.
SARICH, V., LOWENSTEIN J.M. & RICHARDSON, B.J.	1982	Phylogenetic relationships of *Thylacinus* (Marsupialia) as reflected in comparative serology. In M. Archer (ed.) *Carnivorous Marsupials* vol. 2, pp. 707-19. Sydney: Roy. zool. Soc. NSW.
Saturday Evening Mercury (Hobart)	1984	21 January, p. 1.
SAUNDERS, J.	1972	Letter to Eric Guiler.
SAWLEY, F.	1980	Letter to Eric Guiler.
SAYLES, J.	1980	Stalking the Tasmanian tiger. *Anim. Kingd.* vol. 82 (6), 35-40.
SCOTT, P.	1965a	Land Settlement. *Atlas of Tasmania*, pp. 43-5. Hobart: Govt Printer.
SCOTT, P.	1965b	Farming. *Atlas of Tasmania*, pp. 58-65, Hobart: Govt Printer.
SHARLAND, M.S.R.	1939	In search of the thylacine. *Proc. Roy. Soc. NSW* 1938/39 (1939), 20-36.
SHARLAND, M.S.R.	1957	In search of the vanished 'tiger'. *People* 3 April, 25-6.
SHARLAND, M.S.R.	1966	*Tasmania.* Sydney: Nelson, Doubleday.
SILVER, S.W.	1874	*Handbook for Australia and New Zealand.* London: Silver & Co.
SIMPSON, G.G.	1941	The affinities of the Borhyaenidae. *Amer. Mus. Novit.* vol. 1118, 1-6.
SKEMP, J.R.	1958	*Tasmania Yesterday and Today.* Melbourne: Macmillan.
SMITH, G.	1909	*A naturalist in Tasmania.* Oxford: Clarendon Press.
SMITH, S.J.	1980	The Tasmanian Tiger - 1980. Hobart: Tas. National Parks Wildl. Serv.
SPICER, J.	1978	Letter to Eric Guiler.
SPRENT, J.F.A.	1971	A new genus and species of Ascaridoid nematode from the marsupial wolf (*Thylacinus cynocephalus*). *Parasitol.* vol. 63, 37-43
SPRENT, J.F.A.	1972	*Cotylascar is thylacini*, a synonym of *Ascaridia columbae. Parasitol.* vool 64, 331-2.
STEAD, T.	1959	Letter to Hallstrom, 1 Sept. Laird Papers. AOT, NS 1143.
STEPHENSON, N.G.	1963	Growth gradients among fossil monotremes and marsupials. *Palaeontol.* vol. 6, 615-24.
STEVENSON, G.	1972	*Sunday Examiner-Express*, 10 June.
STEPHENSON, R.	1930	In Animals & Birds Prot. Bd. Files, AOT AA 592.
SUMMERS, R.G.	1937	Animals and Birds Protection Board file H/60/34.
SWAINSON, W.	1864	In H. Murray, *Encyclopedia of Geography.* London: Longmans, Green.
TATE, G.H.H.	1947	On the anatomy and classification of the Dasyuridae (Marsupialia). *Bull. Amer. Mus. Nat. Hist.* vol. 88, 97-156.
TEMMINCK, C.J.	1827	*Monographies de Mammalogie.* Tome 1, (XXIII) 264, 23, 60. Dufour 7 c'Ccagre, Paris.
TERRY, E.V.	1961	Letter to Eric Guiler.
THOMAS, O.	1888	*Catalogue of the Marsupialia and Monotremata in the collection of the British Museum (Natural History).* London: British Museum.
The Times (London)	1979	Naturalists hunt extinct tiger in Tasmania. 16 November.
TROUGHTON, E. le G.	1941	*Furred Animals of Australia.* Sydney: Angus & Robertson.
TYNDALE-BISCOE, C.H.	1973	*Life of Marsupials.* London: Arnold.
Van DEUSEN, H.M.	1963	First New Guinea record of *Thylacinus. J. Mammal.* 44, 279-80.
VECHTMANN, N.	1980	Hoe'uitgestorven' is Tasmaanse buidelwolf? *Het Vrije Volk* 13 June.
VRYDAGH, J.M., CARAM, M. & PETTER, F.	1964	*Des fossiles de demain. Treize mammifères menacés d'extinction.* Brussels: Pro Natura Union Int. Prot. Nature.
WALSH, H.	1979	Letter to Eric Guiler.
WARLOW, W.	1833	Systematically arranged catalogue of the mammals and birds belonging to the Museum of the Asiatic Society, Calcutta. *J. Asiatic Soc. Bengal* vol. 2, 97-100.

Washington Post	1980	Tasmanian tiger sought in Australia. 29 May.
Weekly Times	1971	8 September, p. 38.
WEDGE, J.H.	1962	*The Diaries of John Hilder Wedge*, 1824-35. G. Crawford et al. (eds). Hobart: Royal Soc. of Tas.
WEST , J.	1852	*History of Tasmania*. Launceston: Dowling.
WHITLEY, G.P.	1973	I remember the thylacine. *Koolewong* vol. 2 (4), 10-1.
WIBER, J.	1977	Letter to Eric Guiler.
WIDOWSON, H.	1829	*The Present State of Van Diemen's Land*. London: Robinson.
WILFORD, J.N.	1980	A new search for the rare tiger. *Intern. Herald Tribune* 6 June.
WILLETT, J.	1927	Article in the *Hobart Mercury* 13.
WOODBURNE, M.O.	1967	The Alcoota Fauna, central Australia: an integrated palaeontological and geological study. *Aust. Bur. Min. Resources, Geol. Geophys. Bull.* 87.
WOODJONES, F.	1921	The status of the dingo. *Trans. Roy. Soc. S. Aust.* vol. 45, 254-63.
WOODJONES, F.	1929	*Man's place among the mammals*. London: Arnold.
WOODS, S.	1980	Letter to Eric Guiler.
WRIGHT, E.P.	1892	Family LXXII - The Dasyures. In *Concise Natural History*. London: Cassells.
WRIGHT, P.	1984	Radio interview. World Today, 20 Sept.
ZUCKERMAN, S.	1953	The breeding season of mammals in captivity. *Proc. zool. Lond.* vol. 122, 827-950.

Index

A

ABORIGINAL (gines) 13-39-40-41-42-44-45-54-61-71-72-84-111-115-150-157-191
ACACIA 147
ADAMSFIELD 135-154-209
ADELAIDE (Lake) 22-212
ADELAIDE (Zoo) 26-169
AFRICA 9-210
ALLEN (Helyey) 157
ALLIGATOR (River) 42
ALSATIAN (Dog) 153
AMERICA (North and South) 11
AMOS (R.) 149
ANNE (Mt) 60
ANTARCTICA 11
ANTIPODES 9-94
ANTWERP 169
APLICE 145
ARNHEM LAND 42-44-59
ARTHUR RIVER 140-168-185-190
ARTHUR'S LAKE 147
ASIA (Great cat of) 9
AUSTRALIA 11-12-18-32-34-39-40-44-59-60-77-79-88-89-133-185-207
AUSTRALIA (South) 29-30-39
AUSTRALIA (Western) 29-30-35-38-39-40-41
AUSTRALIA (Western A. Museum) 32-34-35-41
AVOCA 195

B

BACKHOUSE 112
BADAJOZ TIER 149
BALFOUR 201-202-204
BANGOR 127
BANKS (Islands) 12
BANKS (Sir Joseph) 13
BARDOT (Brigitte) 186-207-208
BARN BLUFF 61
BARREN (Cape) 85
BARROW (Mt) 126
BART (Mt) 175
BASS STRAIT 59-79-89
BATTY (Wilf) 22-54- 45-152-159-195
BAUDIN (Nicolas) 79
BEAUMARIS (Zoo) 54-133-157-168-169-172-173-174-175-176
BELFAST (Zoo) 177
BELVOIR (Plains) 60
BEN LOMOND (Range) 61-65-126-183-193
BERKITT (Professor) 177
BERLIN 169
BERMUDA 136
BICHENO 126-150
BINKS 152
BISCHOFF (Mt) 151
BLACK BOBS 143

249

BLACK BLUFF 151
BLACKBOY (Plains) 160
BLACKLOW (Jim) 193-194
BLESSINGTON 118-145
BOTANY BAY 78
BOTHWELL 149
BEAM CREEK 127
BRIGGS (Allen) 144
BRISBANE 190
BROADMARSH 182-192-193
BROWN (Dr R.) 193-205-206
BRUNY (Island) 45
BRYANT (J.) 128
BUBBS HILL 186
BUCKLAND 132-144
BUFFALO (H.M.S.) 89
BUNCE 116
BURBURY (F.) 118-132
BURNIE 65-114
BURRUP (Peninsula) 40
BUSBY (P.) 196

C

CADELL (River) 42
CALDER'S PASS 186
CAMERON (Mr) 112-114
CAMERON WEST (Mt) 22-130-131-153
CAMPANIA 132
CAMPANIA HOUSE 154
CAMPBELL (Island) 80-82-83
CARDIGAN (River) 185
CANIS 55
CETHANA (Lake) 72
CHAPLIN (Mr) 176
CHAPMAN 172
CHRYSOCYON 55
CHURCHILL (Elias) 135-136
CLARENCE (River) 145
CLARKE (Island) 85
CLUAN 126
COAL (River) 154
COCKLE (Bay and Creek) 144

COLIN MC KENZIE (Sanctuary) 169
COLLINGWOOD (River) 187
COLOGNE 169
CONDOR (Ship) 212
COONEY (J.) 128
CORINNA 45-191
COTTON (T.) 149
COTTON (W.C.) 158-159
COUNTRY (Lake) 164
COX'S BIGHT 61
CRADLE MOUNTAIN (National Park) 60-61-65-70-125-141
CRANBROOK 126-149-150
CRAWFORD (Dr I.M.) 41
CROWTHER (Dr) 132
CUBE ROCK 153
CUMPSTOW (J.S.) 85
CUON 55
CUVIER (Baron) 88

D

DAMPIER 38-40
DARWIN 17
DASYURIDAE (Family) 11-20
DASYUROPS 55
DASYURUS (Maculatus & Vivervinus) 11-13-14-19-20-55-57
DAVIES (Mr) 119-195
DAVIS (D.D.) 149
DAVIS (Rex) 203-204
DE WITT (Island) 140
DEATH ADDER (Creek) 42
DEE BRIDGE 126
DELORAINE 22-126-136-147
DERWENT (Valley) 126-143-174-186-192
DEVIL'S (Lair) 29
DEVONPORT 65
DIDELPHIS (Cynocephalus) 13-15
DISNEY (Studios) 190
DOHERTY 133
DOLLY 226
DONALDSON (Mr) 192-204

DONALDSON (River) 185
DOVE (Lake) 70
DUNALLEY 77-127-144-145
DUNBABIN (Mr T.) 127-132-144-145
DUNMORE (John) 82

E

EDINBOROUGH 224
ELDON BLUFF 67
ELEPHANTS BAY 79
ELISABETH II (Queen) 172
EMU BAY 112
ENTRECASTEAUX (Bruni d') 77-78
EPPING FOREST 112
ETHEL (Ship) 174
EUCLA (Western Australia) 30
EUROPE 167
EVANS 83-84
EXETER 12
EYRE (Highway) 30

F

FALKLAND (Islands) 80
FINGAL 119-121-160-194-195
FITZGERALD 157-159
FLAMMEA (F.) 213
FLEAY (David) 157-186-187-190-208
FLEMING (Trooper A.) 185-186-187-194
FLORENTINE 136-168-169-193
FLOWER (W.H.) 47
FLYNN (T.T. Professor) 133-140-174-202
FORESTIER (Peninsula) 127-144
FORTH 151
FRANCE 78-80
FRANKLIN (Valley) 67
FRANKLIN (River) 185
FRANKLIN (Square) 167
FRENCHMAN'S CAP 185-200

FREYCINET (Rose & Louis Claude de Saulces de) *79-80*

G

GEOFFROY SAINT-HILAIRE *13*
GEORGE'S BAY *111*
GILRUTH (*Mount*) *42*
GLADSTONE *126*
GLEESON *22-174*
GOLDEN VALLEY *22*
GOULD (John) *87-97*
GRANVILLE (*Harbour*) *200-202-205-212*
GREAT LAKE *132*
GREEN'S CREEK *201*
GRIFFITHS (Jeremy) *204-205-207-208*
GRIM (*Cape*) *112-196*
GUILER (Eric Dr) *186-195-199-200-201-203-207*
GUILER & HEDDLE *51*
GUIMER *136*
GUNN *27-85-146*
GUNNING (G.W.) *154*

H

HEALESVILLE *169-187*
HALLSTROM (Sir Edward) *186-189-190-199-215*
HAMERSLEY (*Ranges*) *42*
HAMILTON *122-125-127-128*
HAMPSHIRE *111-112-116*
HAMPTON (*Tableland*) *30*
HANLON (*Sergeant Inspector George*) *186-191-192-195-199-201*
HARCUS (*River*) *130-196*
HARDING (J.) *128*
HARRIET-SCOTT *98*
HARMON (Ken) *201*
HARRIS (G.P.) *125*
HARRIS (George) *12-13-15-51-79-87*
HARRISON (J.) *133-158-169-173-174*

HASSELBURGH (*Captain Frederick*) *83*
HEAN (Alec) *126*
HELLYER GORGE *159*
HILLARY (Sir Edmund) *186-195*
HIMALAYA *9*
HOBART *8-65-75-77-79-114-125-132-133-147-154-157-167-168-169-172-174-175-176-186-193-196-199-212*
HOOPER (R.P.) *199-201*
HORN (*Cape*) *79-82*
HUBBARD NEWS *190*
HUGHES *19*
HUME (T.) *136*
HUNTER (*Islands*) *79*
HUON (*River*) *78*

J

JANE (*River*) *140-186-187*
JENKINS *126-138*
JERICHO *164*
JERUSALEM (*Walls of*) *154*
JOHNS (George) *147*
JORDAN (*Aide*) *133-136-189*
JOSEPH (E.) *174*

K

KAKADU (*National Park*) *42*
KEAST *25-53-56-57*
KELVEDON *136-149*
KELVEDON (*Estate*) *121*
KIMBERLEY (*Mountains*) *38-41*
KING (*Island*) *79*
KING WILLIAM SADDLE (*Lake*) *127-144*
KNOPWOOD (*Rev.R.*) *77-78-79*
KREFFT *12-14-15-17-56*
KRETSCHNER *98*
KUHNERT *103*

L

LADY NELSON (*Ship*) *89*

LAIRD (N.) *190-215*
LANDS (*Department*) *20-125-127-128-143*
LANKASTER *108*
LAPÉROUSE *78*
LAUNCESTON *51-65-75-80-133-134-136-169-175-197*
LE SOUEF *57-169*
LEAKE (*Lake*) *149*
LEFEVRE (John) *159*
LEMONT *149*
LEVEN (*Canyon*) *72*
LEWIN (John William) *39*
LIFE (*Magazine*) *190*
LINNEAN (*Society*) *13-39*
LISDILLON *126-149*
LISTER PARKER (Mr W.) *39*
LITTLE SWANPORT *126*
LOCH NESS (*Monster*) *9*
LOFTY RANGES *135*
LONDON *27-79-85-89-133-167-169-172-173-175*
LORD (C.E.) *19-150*
LOUIS XVI (*King of France*) *78*
LOWRY (David & Jacky) *29-30-31-35*
LYDEKKER (Richard) *105-109*
LYELL (*Highway*) *143-186-200-210*
LYNE (J.) *119-121-125*

M

MAC INTYRE (*Inspector*) *193-194*
MACQUARIE (*Harbour and Island*) *83-125-152-204*
MADURA *30*
MAGALA (*Creek*) *42*
MAGARNI *42*
MAINWARING (*Inlet & River*) *147-210*
MALAHIDE (*Estate*) *121-195*
MALLEY (James) *23-130-137-186-195-204-205-206-207-208*
MAMMOTH (*Cave*) *29*

MANN (River) 42
MARBLE BAR (Area, Western Australia) 41
MARIA ISLAND 141-202
MARSUPIAL (Wolf) 15
MARTIN (Ray) 201
MAWBANNA 22-136-143-152
MAYFIELD 149
MC KAY 112
MC KAY'S (Property) 47
MEERDING (Hank) 200
MELBOURNE 169-177-199
MENDOCA (Pedro de) 77
MEREDITH (Louisa Ann) 21-146-147
MERRILEES (Durcan) 32
MIDDLESEX (Plains) 185
MIDLANDS (Tiger Association) 122
MILABEENA 157
MILOTTE (Al & Elma) 190
MILTON (F.) 157
MINI (Art) 42
MIOCENE (Period) 29
MOELLER (Heinz) 25-47-48-54-55-56-167-169
MOLE (Creek) 126-212
MONARCH (Ship) 174
MONTUMANA 136
MOONEY-TOGARI (Expedition) 186
MOORE (T.B.) 150-152
MOORE'S VALLEY 152
MORRIS 199
MORRISON (B.) 198-199-200
MORRISON (Mr) 118
MOULTING (Lagoon) 158
MT FIELD (National Park) 75-193
MUDIE 84
MULLINS 175-176
MUNDRABILLA (Station) 30
MURPHY (Elain) 159
MURRA-EL-ELEVYN (Cave) 33
MURRAY (Miss) 160-161

MUTZEL 98

N

NAARDING (Ranger H.) 210-211-212-215
NABARLEK 211-215
NAMALAWU 42
NATIONAL ZOOLOGICAL PARK 22
NATIVE CORNERS 132
NEW GUINEA 11
NEW HARBOUR 61
NEW NORFOLK 174-193
NEW SOUTH WALES 39-78-83-173-199-204
NEW YORK 9-24-169-174
NEW ZEALAND 79-80-88-103-159
NEWCASTLE 173
NORTHERN IRELAND 177
NORTHERN TERRITORY 44-59
NUBEENA 127
NULLARBOR 29-31-32-33-34-35

O

OATLANDS 122-132-149
OBIRI (Rocks) 42-44
OENPILLI 44
OLD HAIRY 32-34-35
OPOSSUM 15
OUSE 143-152
OXLEY 77-78-79

P

PADMAN (W.) 118
PAPUA NEW GUINEA 29
PARATTAH 132-148
PARIS 9-169
PATERSON (Lieutnant Governor) 77-78-79
PEARCE (D.C.) 172
PEARCE (H.) 20-21-25-26-127-143-144-145-147-154
PEARCE (T.) 116-118-126-138-143
PEARCE (W.) 192
PEARSON (D.J.) 185
PELVERATA 165
PENNY (C.) 150
PERTH 35
PIEMAN (River) 27-136-140-185-189-190
PILBARA 41
POCOCK 50
PONY BOTTOM 144
PORT ARTHUR 83
PORT DALRYMPLE 78
PORT DAVEY 127-204
PORT JACKSON 79-80-82-83
POURRAT 88
PYENGANA 136

Q

QUAKER 12
QUEEN VICTORIA (Museum) 17-133-169-197
QUEENSLAND 39
QUEENSTOWN 152

R

RAGGED TIER 127-137-145
RAGLAN (Range) 185-200
RASSELAS 136
REUBEN (Charles) 203
RECHERCHE (Bay) 77
RED BANKS 149
REECE (Premier) 191-201-202
REYNOLDS (H.) 127
RINGAROOMA 148
RIO DE JANEIRO 9
ROBERTS (Mr) 133
ROBERTS (Mrs) 157
ROBINSON (George) 45-72-111-157
ROLAND (Mt) 67

ROSLIN (Institute) 224
ROSS 125-128-132-149
ROSSARDEN 75-186-189-193-195
ROWE 27-169
RUSSEL FALLS 75

S

SACK (Mrs M.) 41
SAFETY COVE 144
SANDSPITT 144
SANDY CAPE 126-157-186-198-199-200
SARCOPHILUS 55
SAWBACK RANGES 209
SAWFORD (W.) 148-149
SAWLEY 23-26-47
SAYLES 186-206
SCHINZ 89
SCHREIBER 102
SCOTT (Harriet) 98
SEYMOUR (District) 157
SHANNON (River) 133
SHARLAND (M.S.R.) 140-141-154-157-172-187
SHAW (F.) 149-150
SHAW (Mr) 121
SINGAPORE (Republic of) 227
SMITH (George) 164-202
SMITH (J.) 151
SMITH (S.J.) 145-182-183-199-200
SMITH (Steve) 207
SMITHSONIAN INSTITUTE 173
SMITHTON 126-130
SPECHT (E.) 103
SPERO (River and bay) 150-165
SPRING BAY (Association) 122
ST CLAIR 125-141-187
ST HELENS 121
ST PETER'S PASS 118-149
STACK'S BLUFF 193
STANNARD 126-138

STEVENSON (George) 159
STEVENSON (R.) 145
STIRLINGSHIRE (Ship) 172
STRAHAN 136-152
STRICKLAND 143
STRONG (Cave) 29
STUDLAND (Bay) 130-132-180-196
SUEZ (Canal) 167
SUMMERS 140-141
SUMMERS (Sergeant) 185-186-202
SURREY HILLS 111-114-115-116
SUTTON (Noel) 194-195
SWAINSON'S (Nation) 54
SWANSEA 119-121-122-126-149
SWANSTON 148
SYDNEY 9-30-80-82-133-160-164-169-173-174-175-177-189-212

T

TAMAR (River) 77
TANGENY 206
TANGEY 186
TANINA 192-193
TARONGA PARK (Zoo) 169-189-199
TASMAN (Abel) 77
TASMAN (Peninsula) 127
TASMAN (Sea) 65
TASMANIA 18-19-21-23-29-38-40-45-47-53-59-60-61-64-67-70-71-75-80-82-83-85-87-88-105-111-115-118-122-125-126-132-134-137-138-146-151-153-157-167-169-172-175-177-181-182-195-197-202-204-206-208-210-211-213
TASMANIA (University of) 17-18-133-140-174-199
TASMANIAN (Devil) 11-19-20-21-25-27-31-39-48-51-52-53-54-55-59-72-115-121-147-180-181-182-187-194-199-200-208-212
TASMANIAN (Dingo) 15
TASMANIAN (Museum) 17-18-19-150-173-175-177-185
TASMANIAN (Wolf) 1-15
TASMANIAN (Tiger) 8-9-11-13-14-15-17-18-19-20-21-22-24-25-26-27-29-31-34-38-39-40-41-42-44-45-47-48-50-52-53-54-55-56-57-59-60-71-75-77-78-80-83-84-85-89-94-105-111-112-114-116-118-119-121-122-125-127-130-132-133-134-135-137-138-139-140-141-143-144-146-147-148-150-154-157-160-164-165-167-168-169-172-173-175-176-177-179-180-182-183-185-187-189-190-191-192-193-195-196-197-198-201-202-203-207-211-212-213-215-224
TEMMA 165-201
TEMMINCK 13-15
TENALGA 175
TERREY (R.) 213
THE DEN 149
THYLACYNE (Hole) 30-32-33
THIRKELL'S CREEK 186
THOMAS (C.) 149
THOMPSON'S MARSHES 159
THOMSON (L.) 198-199-200
THREE STICKS (Run) 130-196
THYLACINUS (Breviceps) 15
THYLACINUS (Cynocephalus) 11-13-14-15-17-31-50-52-55-58-64
TIERS (Eastern & Western) 51-125-135-147-149-212
TOGARI 210-215
TOOMS (Lake) 118-121-149
TOP FARM 205
TOTTENHAM COURT (Road) 172
TOULON 80
TRAVELLER'S RANGE 136
TRIABUNNA 121
TROWUTTA 23-47-136-159-183-

186-195-204-212
TURNBULL 189
TURNER (Ted) 212
TYENNA 127-175

U

UNITED KINGDOM 140
UNITED STATES 60-190
URANIE (Ship) 79-80-83

V

VALENTINE'S PEAK 195
VAN DIEMEN'S (Land) 8-78-79-82-84-111-112-114-116-121-122-125-137-150
VICTORIA (State) 29-39-59-169-187

W

WAINWRIGHT (George) 22-114-130-132-134-138-168
WALKER 192-193
WALSH (Harry) 165
WANDERER (River) 213
WARATAH 125-128-150-169
WARDE (E.) 128
WARSDEW'S LIVERY STABLES 175
WASHINGTON 22-169-172-173
WATERHOUSE 92
WEEKLY COURRIER 150
WELCOME (Heaths and River) 130-196
WERRIBEE 169
WHALE'S HEAD 165
WHYTE RIVER 186-191-203-204
WIDOWSON 84
WIGG (P.) 192
WILL (Lake) 61
WILLETT 146-149
WILLIAMS 185
WILMOT (Power station) 72

WILMOT (Sir John Eardley) 167
WILMOT (Zoo) 72-146
WILMUT (Ian Professor) 224
WOODS (S.) 148-151
WOOLNORTH 22-111-112-115-116-122-125-126-128-130-131-132-134-135-137-138-165-168-169-180-181-186-196-197-201-202-215
WORLD WILDLIFE FUND 202-207
WRIGHT 186
WRIGHT (E. Perceval) 94-105
WRIGHT (P.) 213-213
WYNYARD 133-158-169

Y

YORKTOWN 77
YOUD (A). 22

Z

ZEEHAN 164

Acknowledgments

Mr B.H. Arnold, *Victoria*
Prof. Gabriel Arvis, *Paris, France*
Mrs Chantal Attali, *Nouméa, New Caledonia*
Dr Alex Baynes, Western Australian Museum, *Perth, Western Australia*
Dr. J.L. Bannister, Western Australian Museum, *Perth, Western Australia*
Mr C. Binks, *Devonport, Tasmania*
Mr Charles Bouvier, *Paris, France*
Dr J. Calaby, *Canberra, A.C.T.*
Mrs Kathryn Davidson, *Hobart, Tasmania*
Mr (+) and Mrs Peter Dombrovskis, *Hobart, Tasmania*
Mr P. Donovan, N.Z.T.V., *Dunedin, New Zealand*
Miss Gina Douglas, *London, Great Britain*
Prof. Dr Heinz F. Moeller, Institut für Zoologie der Universität, *Heidelberg, Germany*
Miss Holly Gibson, *Perth, Western Australia*
Miss Cécile Godard, *Darlington, Western Australia*
Mr Emmanuel Godard, *Nouméa, New Caledonia*
Prof. Alex Kahn, Institut Cochin de Génétique Moléculaire, *Paris, France*
Dr Philippe Kourilsky, Institut Pasteur, *Paris, France*
Dr John Long, Western Australian Museum, *Perth, Western Australia*
Mr David C. Lowry and Mrs W. Jacoba Lowry, *Linden Park, South Australia*
Mr Kenneth McNamara, Western Australian Museum, *Perth, Western Australia*
Prof. William Marsden, *Brisbane, Queensland*
Mr Christian Meier, *Herford, Germany*
Mrs Fiona Mowat, Collections Manager, Aboriginal Affairs, Museum & Art Gallery of the Northern Territory, *Darwin, Australia*
Mr Didier Murcia, *Perth, Western Australia*
Prof. Svante Pääbo, *Zoologisches Institut, München, Germany*

Mr J. Reynolds, *Yolla, Tasmania*
Mr Alex Risco, *Perth, Western Australia*
Mr G. Roberts, *Launceston, Tasmania*
Mr Nicolas Sévenet, *Paris, France*
Dr V. Smith, University of Sydney, *New South Wales*
Mr C. Tassell, Quenn Victoria Museum, *Launceston, Tasmania*
Mrs Pat Vinnicombe, Policy Officer in the Aboriginal Affairs Department, *Western Australia*
Mr R. Wheeldon, University of Tasmania
Mr Gerard Willems, *Hobart, Tasmania*
Mr C. Williams, University of Tasmania
Mrs Priscilla Wright, *Balga, Western Australia*

Aboriginal Community of the Pilbara, *Western Australia*
Archives of Tasmania, *Hobart, Tasmania*
National Parks and Wildlife Service, *Tasmania*
National Zoological Parks, Smithsonian Institution, *Washington D.C., USA*
Queen Victoria Museum, *Launceston, Tasmania*
Tasmanian Museum and Art Gallery, *Hobart, Tasmania*
The Editor, Zeitschrifte für Zoologische Systematik und Evolution Forschung
The Linnean Society of London, *Great Britain*
Western Australian Museum, *Perth, Western Australia*
World Wildlife Fund Australia

Mr N.J. Mooney discussed several sections of the text with Eric Guiler and the late Lalage Guiler, the author's wife, helped with the early editing.

All those people who helped in the field investigations deserve special thanks, particularly the last surviving members of the 1963 Expedition, R.G. Hooper and K. Harmon. Thanks also to those people who provided details of sightings of this elusive animal.